The Humbled Blue Orb

The Humbled Blue Orb

◆

A journey through concepts in astronomy,
cosmology, relativity, quantum physics,
particle physics and string theories

David Klein

iUniverse, Inc.
New York Lincoln Shanghai

The Humbled Blue Orb

A journey through concepts in astronomy, cosmology, relativity, quantum physics, particle physics and string theories

iUniverse books may be ordered through booksellers or by contacting:

iUniverse
2021 Pine Lake Road, Suite 100
Lincoln, NE 68512
www.iuniverse.com
1-800-Authors (1-800-288-4677)

All illustrations were created by the author.

Specific screen-shot credits are provided in the text where applicable.

ISBN-13: 978-0-595-38706-9 (pbk)
ISBN-13: 978-0-595-83088-6 (ebk)
ISBN-10: 0-595-38706-3 (pbk)
ISBN-10: 0-595-83088-9 (ebk)

Printed in the United States of America

For Sam:
May your curiosity be unbound.

Contents

Introduction

My interest in astronomy began in 1986 when Halley's Comet was making its most recent appearance. My neighbor across the street had been into telescope making and amateur astronomy for a number of years. And, being a teacher by profession, he was kind enough to show me the ropes and pretty soon my dad and I were ordering optics and making measurements for my first homemade telescope. I was in awe. To be able to see first hand all of these distant objects that I had read about introduced a humbleness that transcends words.

Despite going on in later life to get a degree in Meteorology, I have always felt astronomy to be one of my deepest passions. While I do not use my telescopes nearly as much as I used to, or should, I look up at the stars every time I am outside at night and take a moment to find a couple of constellations or planets and think about just how small we really are on our little blue orb. When you look up at the stars you are in essence looking at a piece of yourself, as we are all made from the same components as those stars (and all matter in the universe for that fact).

The purpose of this book is to capture some of the magic that is intrinsic in physics, astronomy and cosmology while explaining some of the most recent developments in the field. We will explore all scales of the cosmos; from the sub-atomic realm, where quantum physics dominate, to the largest structures in the universe, where Einstein's relativity become significant, and black holes and the first instants of the big bang, where both theories must be combined into a higher theory which may come to fruition in m-theory and superstring theory. Everything in our world is on its way to somewhere or something else, and the fields of physics, astronomy and cosmology are no exception. The speed of book publication and software development has no hope of keeping up with the daily developments in scientific fields. While information changes rapidly, concepts do not. If nothing else, I hope you come away from reading this book with a good sense of the concepts of astronomy, cosmology, and modern physics so that when the inevitability of change rears its head and throws in new information, you can interpret that new information with a good conceptual foundation.

To aid in establishing this foundation I also have examined and identified many freeware software packages that will enable the reader to download and experiment with some of the data and concepts that are discussed. These programs provide a hands-on approach that every curious reader will profit from. With some of these programs, such as SETI@Home and Einstein@Home you can use your personal computer to aid in scientific research efforts. I personally find these efforts very exciting. I have been searching for extra-terrestrial intelligence via SETI@Home for over two years now, and have recently started searching for gravitational waves from spinning neutron stars (pulsars) with Einstein@Home.

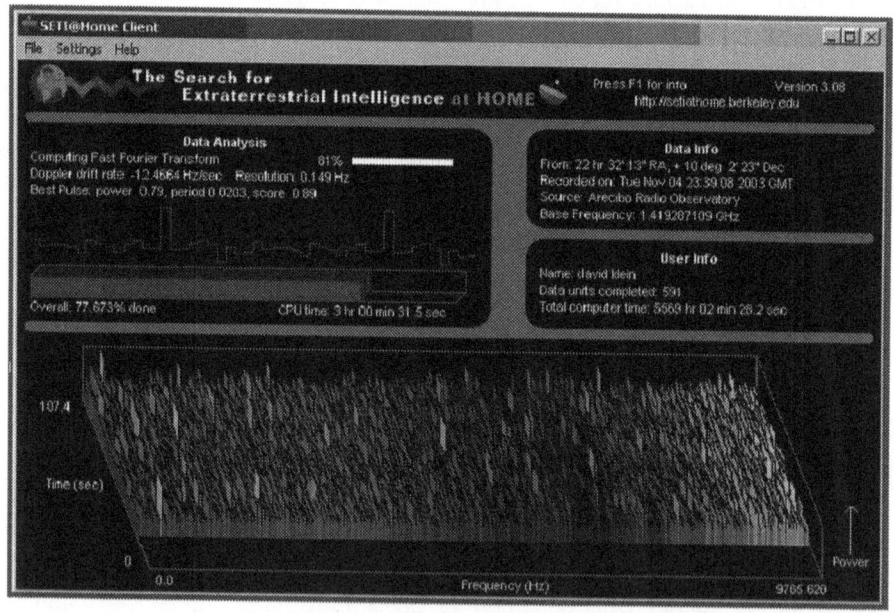

Screen shot of the SETI@Home screen saver.

One last note before we delve into our journey. Science is an evolving realm that requires a mind that is open to new ideas. I have attempted to re-create to the best of my ability and knowledge the current concepts of astronomy, cosmology and physics. As with any type of knowledge, you need to continue to feed and nourish it for growth. Many of the ideas presented will be counter-intuitive and may take time to fully grasp. Your mind may need to warm up to these ideas over time, allowing more malleability, before wrapping around certain topics. I hope that this book will allow you to have a greater understanding of our cosmos that leaves you with a sense of amazement, and a thirst for advancing your knowledge beyond these pages.

"I have been like a boy, playing on the seashore, diverting myself and now and then finding a smoother pebble or prettier shell than usual, while the great ocean of truth lay before me, all undiscovered."

—Newton

1

Scaling up…

The purpose of this book is to explore the mystery and elegance that abounds our cosmos. Astronomy and cosmology are fields that cover a vast range of scales, from the subatomic quantum field, to the largest structures observable in the universe. Scale is perhaps a good thing to establish first off, as it is a very humbling realization when dealing with astronomical and cosmological distances. In our everyday lives, we are greatly limited to the scales that we are able to experience as we live on a planet. Unless you are an astronaut, a patient pilot, or an avid balloonist, the largest distance you can hope to experience personally is the circumference of the earth at about 25,000 miles. Even that distance, the largest available on our planet, has only been experienced by a minority of the human population.

In astronomy, the scales involved are hard for the mind to grasp due to our spatial limitations. Humans are experience driven, and it is hard to imagine any sort of distance that starts getting into the millions, billions, and trillions of distance units. Our only hope of experiencing such immense numbers when dealing with our distances would be to start measuring using units such as the micron or nanometer on the way to the office. Even the distance from the earth to the sun is hard to imagine. This distance compared to the cosmic whole that we currently know is so slight, that I doubt if humans even have the precision to express it.

According to Einstein's theory on Relativity the speed limit for the entire universe is the speed of light in a vacuum, which is about 186,000 miles per *second* (or about 300,000 kilometers per second). This speed limit is enforced by changes in mass that are linked to an object's velocity. More specifically, as velocity increases, so does mass. As any object *with mass* approaches the speed of light, the mass increases to such a degree that the energy required to propel it becomes more and more (approaching infinity). Actually, *any* object with mass, even the smallest subatomic particles are impossible to accelerate to the exact speed of light. Because of this, technically, when you are flying on an airplane, or driving in your car on the way to work, you have more mass (if the person determining your mass is not in the car or airplane with you and is outside your frame of reference). It should also be noted that when you are in an airplane, the effects of being at a high elevation, away from the earth's center of gravity *decreases* your mass many orders of magnitude more than the velocity of the aircraft increases your mass due to the effects of relativity. With the velocity mass increase you must remember that you are only slightly more massive as seen by someone not in the plane. In the unmoving frame of reference inside the plane, your mass would appear unchanged. The only reason we do not notice this is that until you reach velocities that are a significant percentage of the velocity of light, they are miniscule. The scales and velocities in our everyday lives where these strange effects can be ignored are what scientists refer to as "classical" scales (technically—relativity is classical as well as it does not incorporate quantum effects in its calculations—however these are so small as to be disregarded for all but the tiniest of scales). When traveling at the relatively slow speeds of a car or an airplane in comparison to that of light, you also do not notice because if you were using a scale inside the airplane or car that was traveling the same speed as you (in your same reference frame) then your mass would appear normal. Only to an *outside* observer that was measuring your mass as you passed by, would this difference be measurable. Everything is relative to your frame of reference that you are measuring from. The same events may appear very differently as viewed from different reference frames moving at different velocities. This concept can be confusing; however it is a key to understanding Relativity. A deeper presentation of relativity will be presented later in this book.

I have been careful in the above section to use the term "mass" and not "weight", as they are two different things. *Weight* is determined by taking your mass and multiplying it by the strength of the gravitational field that the object is in. *Mass* is an intrinsic value that reflects the amount of matter that an object is composed of. An astronaut on the moon will *weigh* less than on earth because of the reduced gravitational field of the moon as compared to the earth; however their *mass* will remain unchanged.

Let us get back to our cosmic "backyard" as it were, and explore the earth-sun distance. After traveling for only one second, a photon of light zipping along has covered a distance that is beyond the reach of most human experience. To place some more perspective on this, let us imagine how long it takes you to drive 186,000 miles in your car. If statistically the average American driver puts around

15,000 miles a year on their car that means it would take about 12.4 *years* of your everyday errands, vacations, and commutes to work to drive the distance that a photon of light goes in one *second!* Keeping this in mind, let us expand the time we let our photon race along from one second to around 8 minutes—and then we would be at the sun. Everything that happens on the sun takes 8 minutes to reach us. If the sun suddenly went black—we would still be able to enjoy 8 minutes of sunlight before we realized anything was wrong.

So these 8 minutes of light-speed velocity has placed us at our sun, a distance that in astronomy is referred to as an Astronomical Unit, or AU. One AU works out to be to be 149,603,500 kilometers, or 92,961,847 miles. In the realm of our solar system this unit of measure is common. If we wish to explore beyond our solar system, we need to adopt a new measuring stick; one that will save us the time of writing out zero's all day. For our 1 AU example above, if you were driving 100 miles per hour it would take you around 106 *years* to cover that same distance.

For interstellar distances—the distances to the nearest stars—we need to start thinking in terms of how far a photon traveling at the speed of light can travel in not seconds, minutes, or hours, but *years*. Practically everyone is familiar with the term light-year these days. Just as it sounds, a light-year is the distance that light travels in one-year. This distance turns out to be around 10 *trillion* kilometers or around 6 *trillion* miles. Even for the nearest star outside of our sun, Proxima Centauri, we are looking at a distance of 4.3 light-years, or 25.8 *trillion miles* (25,800,000,000,000 miles). It takes 106 years to travel one AU at a speed of 100 miles per hour. To get to Proxima Centauri at the same velocity would take 28,620,000 years! If you could travel and maintain speeds of say, 20,000 miles per hour, it would still take about 147,260 years to get to there—and that is the *nearest* star in a "typical" galaxy that contains *billions* of other stars.

As you may have noticed, the scale increase from dealing with solar system distances to interstellar distances is not linear; it is many orders of magnitude larger, or exponentially larger. This type of scaling happens often in astronomy, and tends to dilute the comprehension of the distances being described when people read about them. This is one of the prime reasons that I chose to start this book with distances and scales—to establish, up front, a sense of awe and respect for the magnitude of volume that our universe encompasses.

Continuing on our scaling journey, let us expand our scope from interstellar distances to galactic distances. For this scale, we can continue to use our light-year distance ruler; however we will need to start thinking in terms of tens of thousands of light years just to capture the distances out to the perimeter of our Milky Way galaxy. Specifically, our Milky Way is about 100,000 light-years in diameter, or 600,000,000,000,000,000 miles.

Before we expand our scale one again, I would like to stop once again and re-iterate the magnitude of the sizes we are dealing with at this level. At the scale of the Milky Way galaxy we are suddenly faced with billions of stars. Each of which is a size that dwarfs our Earth. Our own sun, a fairly typical star out of the *billions* in our Milky Way is large enough in volume to contain *1 million* Earth-sized planets. Even Jupiter, at a fraction of the size of our sun, is large enough that our entire planet could fit inside it's famous raging maelstrom, the Great Red Spot, which itself is only a small fraction of the planet's volume.

Our next leap is to the scale of a cluster of galaxies that the Milky Way belongs to that is known as the Local Group. To explore this swath of real estate we are going to need to cover *millions* of light years. Galaxy clusters are relatively common at the scales of millions of light years. In addition to our Local Group, there are many clusters of galaxies, such as the Virgo, Hercules and Coma clusters where the density of galaxies is enough for their gravitational attraction to over-come the overall expansion of the universe. The Andromeda Galaxy, a "neighbor-ing" galaxy to our Milky Way, is 2 *million* light-years away, or 12,000,000,000,000,000,000 miles. Our Local Group and its 20 galaxies are a very small clustering, and even they are spread out over a distance of 3.3 *million* light-years, or 19,800,000,000,000,000,000 miles.

Despite the colossal scales at the level of galaxy clusters, we still have more ter-rain to cover. Nature has sowed her seeds farther. If we expand out to *hundreds* of *millions* of light years we will see huge filaments and shells of structure that are millions to hundreds of millions of light years across. Each of these structures is made of *thousands* of galaxies that each contain *billions* of stars like our own sun that can house a *million* earth's within them (actually—there is quite a large spec-trum of stellar sizes in the stellar range from the smallest black holes to the largest super giants. On average, you could probably safely assume the average star size would house around a million earth size objects.

Intriguingly, at this level, there are also huge voids that the above mentioned arcs and shells seem to reside on the perimeter of. The patterns are akin to throwing a handful of sand grains onto an over-turned egg carton, where the oval "bubbles" of the egg-carton force the grains of sand to settle on their edges. There are a few grains that are lucky enough to stick to the material of the egg-carton, catch a small deformity, or fall close enough to the center so as to not have gravity pull them off the slope of a shell; but overall they are forced to the edges of the inverted shells that keep the eggs in place (another good example of this type of pattern is to look at the inside of a slice of very fibrous bread).

This is the level of the largest known structures in the universe. One of these great arcs of galaxies is known as the Great Wall, and extends for hundreds of millions of light-years. These arcs and voids are very likely to extend all the way out to beyond *billions* of light years—to the edges of the observable universe (any reader who has an interest in visually exploring these structures first hand will greatly profit from downloading a freeware software package named Galaxy Explorer, which as of this book's printing was available at the SDSS website at: http://cas.sdss.org/dr3/en/help/download/ (SDSS is the Sloan Digital Sky Survey). This software allows for a virtual fly-through of 30,000 galaxies of the SDSS from any angle you desire. It is very informative and powerful tool to visualize the largest structures currently visible in our universe. Below is a screen shot (colors are inversed) from the Galaxy Explorer software that readily shows the great arcs and voids of galaxies:

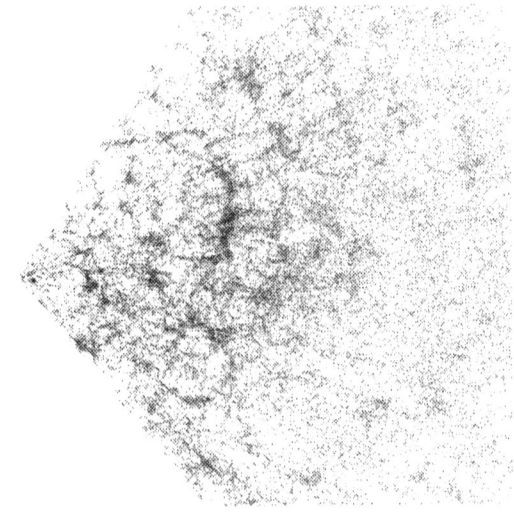

These represent the largest of the known structures in the cosmos, with each of these arcs consisting of thousands upon thousands of galaxies that within themselves contain billions of stars. The scales here are truly staggering.

Once we start encountering billions of light years of space we are finally getting to the end of our scaling journey. For at this level we are reaching the edges of our "visible" universe. According to the most main-stream big bang theory (we will cover more in-depth later) time-line, the entire universe is somewhere around 13.7 *billion* years old. Again, 13.7 billion years is an amount of time that is incomprehensible to a human. The longest timescales that humans are capable of experiencing are their lifetime (let's say about 80 years). Compared to the approximately 13.7 billion year old universe, this would constitute a 0.0000005714285714 percent (or 5.7142857142857100E-09% if you prefer scientific notation) slice of the whole 13.7 billions years of our cosmos.

I placed the word "visible" above in quotations since more recent measurements of the expansion of the universe indicate that it is in fact *accelerating* as it is expanding. This has a significant implication that cosmic structures that are beyond a certain distance will never be observable from our location as the light emitted by them will never be able to overtake the accelerating expansion of the universe at such an early epoch (again—we will cover this terrain later. Astronomers and cosmologists like to refer to these types of situations as "event horizons" since, just as with a real horizon, unless you are able to move the point from which you are making your observations, you will never be able to obtain any *direct* information from beyond that horizon distance due to the curvature of the earth).

Some people think that since the universe is estimated at 13.7 billion years old, and since nothing can travel faster than the speed of light, that the cosmos most be 13.7 billion light years in radius. This is in fact *incorrect* since the fabric of space-time *itself* is expanding. The farther and farther you get from any point in our cosmos, the more and more expansion of space-time there is between those points, and hence more distance that needs to be traveled, and hence more *time* that must go by. Based on this insight, the *radius* of the *observable* cosmos is around 45 *billion light-years* for a diameter of *90 billion light years*. In miles, 90 billion light-years (with 1 light-year equaling about 6 *trillion* miles) would be about 5.4×10^{23} miles or 540,000,000,000,000,000,000,000 miles! To further humble this number, remember from earlier that the average American driver puts around 15,000 miles a year on their car, and that it would take 12.4 years to travel the distance of 186,000 miles that light travels in one second. So how long would it take for an average American driver to cover the distance of the *observable* cosmos? It would take about 2.11×10^{32} years. That is about 10^{22} times longer than the 13.7 billion year old universe has existed.

Distances of these magnitudes become philosophical. It is difficult for a mind to wrap itself around such scales that we have covered in this chapter. As with many other types of situations that we encounter in our everyday lives, I feel that it is wise to step back and see the whole picture before delving into any particular part of it. That is the main reason for choosing distances as a starting point in our journey. (Later in this book we will have a chapter to "scale down" our sense of distances, and reorient our mental picture as we probe down to the subatomic and quantum levels of our universe).

There is a nearly infinite playground in our universe for us as a species to ponder and explore. I feel that people must allow themselves the opportunity to break from their routines, and really think about how small our daily reality is when observed next to the scales around us in the cosmos. The reality is that even if you somehow dumped all of the sand from all the beaches on earth into one enormous pile, and called that pile our cosmos, we *might* be lucky enough to be classified as *one* grain of sand from that pile. However, running to our rescue from this very humbling analogy is the fact of our intelligence. Humans may only occupy one of those grains of sand; however we are the most intelligent grain that we know of in the pile, and that adds an overwhelming importance to our existence. Until we discover other intelligence, we are quite literally the conscious universe (in a recent PBS survey of whether people believe there is intelligent life in our own galaxy, 78% indicated that they believe that intelligent life resides in our own galaxy, while 15% disagreed with the notion, while 7% were undecided. At the time that I found this survey on www.pbs.org these results were based on 11,731 accumulated votes as of September 2005).

2

The Big Bang

In the previous chapter we established a sense of scales and distances that are routinely used in astronomy and cosmology. I ended with an emphasis on the difficulty in the human mind to comprehend scales such as light years when it would take around 12.4 years to travel as far as a light does in one second. Well, brace yourself, because the most successful cosmic theory in history is even more difficult to wrap your head around—the big bang.

First off, let me start by saying that there have been many different theoretical models of the universe over the course of cosmology. I am only going to explain the basic Big Bang model as it represents by far the most widely accepted model with a substantial amount of evidence in its favor (it is more of a fact now than a theory in certain circles). Most modern cosmological models are based on the big bang, or variations on the basic ideas presented in it. The two biggest pieces of evidence that placed the Big Bang theory into the limelight are the discovery of the expansion of the universe, and the discovery of the remnants of the big bang itself, the cosmic background radiation.

There are some common misconceptions that seem to follow the interpretation of the Big Bang model that need to be clarified up front. The name Big *Bang* tends to imply an explosion that we are used to seeing on television and in movies, and this is simply not the case. You must unlearn a lot of your everyday intuitions and open your mind. This event did not occur on a planet with a localized source of gravity in a universe that is 13.7 billion years into its evolution—it occurred from nothing. It is not correct to visualize the Big Bang as an explosion at a point. The three space dimensions and the time dimension that make up our current universe were not in existence yet prior to the Big Bang—they were *created* in the initial expansion of the Big Bang. A more accurate way to think about it is that in the beginning of the universe *everything* was expanding at nearly the speed of light from *all* directions—not from a point. If you could somehow go back in time 13.7 billion years and see the Big Bang, you could not isolate it to a point in space where it began, because the "space" (space-time) that is needed to isolate it to a point is what is being created *with* the expansion itself.

It is not accurate to think of the start of the Big Bang universe as occurring at a single point. Rather, it is more correct to think of our *observable* universe (as far out as we can ever hope to see with our technologies both now and in the future) as being packed down into something incredible dense of an *arbitrary* size—say the diameter of a golf-ball. The reason to think of something arbitrary in size is because the universe may in fact be infinite due to the intrinsic distance limitation of what we are able to observe from earth. You can shrink down something that is infinite in size for an infinitely long time and still have an infinite universe, thereby making the size of the universe after the first instants of the big bang arbitrary.

Yet another reason we have no reason to believe that the universe according to the Big Bang started as a point is because the mathematics of the model break down and start throwing out infinite values at the smallest of scales. When formulas start spitting out infinities it is does not necessarily *literally* mean that a value (such as density or volume…etc) is infinite—it can also mean that the mathematics used are not structured in a way that is capable of modeling a situation anymore. The big bang model and its embedded mathematics provide an exceptional model for explaining scientifically the observed properties of our cosmos, but it has no way of explaining anything *before* the "bang". Science needs a new kind of theory and mathematics that can transcend this boundary. Right now, that does not exist (there are some hopeful theories that may provide this—such as string theory and m-theory which requires a whopping 10 and 11 space-time dimensions for the universe to work right—this will be covered later in this book).

People are often needlessly confused by the term space-time and how space is four dimensional. Technically, a dimension is nothing more than a unique degree of freedom. If you have an apple in your hand, you can move it up or down (one dimension), or left or right (second dimension) or towards you or away from you (third dimension). But whichever way you move the apple, it is in a unique place in *time* (fourth dimension). It always takes a certain length of time to move the apple, even if you are moving the apple at the speed of light—the fastest speed possible in the cosmos. Therefore time is another unique degree of freedom that is interwoven into the three spatial dimensions.

Another example of our four dimensional world is to imagine you are meeting a business colleague for a reserved lunch in a two-storey restaurant downtown. If you want him to show up, you need to give him the address of the restaurant (x-y coordinates where x = one dimension, y = two dimension), which level your table is reserved on (z coordinate = third dimension), and perhaps most importantly, what *time* you are going to be there (fourth dimension), otherwise he will just show up at some arbitrary time which may or may not correspond to the time that you show up to meet him.

Another misconception of the big bang is the effect of the expansion of the universe has on space. This cosmic expansion in fact has been found to be an *accelerating* expansion. A perfect analogy for visualizing the expansion of our universe is to inflate a balloon to a small diameter and then place a bunch of black dots all over it with a marker. Now blow up the balloon to successively larger and larger diameters and observe what happens to your black dots. While the surface of the balloon is expanding and increasing the distances between your dots, the dots *themselves* are the relatively unaffected (in a real scenario where the "dots" represent galaxies and clusters of galaxies, these structures are expanding with everything else—however their localized gravity overcomes this expansion and maintains their sizes). If you imagine our Milky Way galaxy and all galaxies and matter in the universe as being on the surface of an enormous balloon (the whole cosmos), that is a more appropriate way to think of the cosmic expansion.

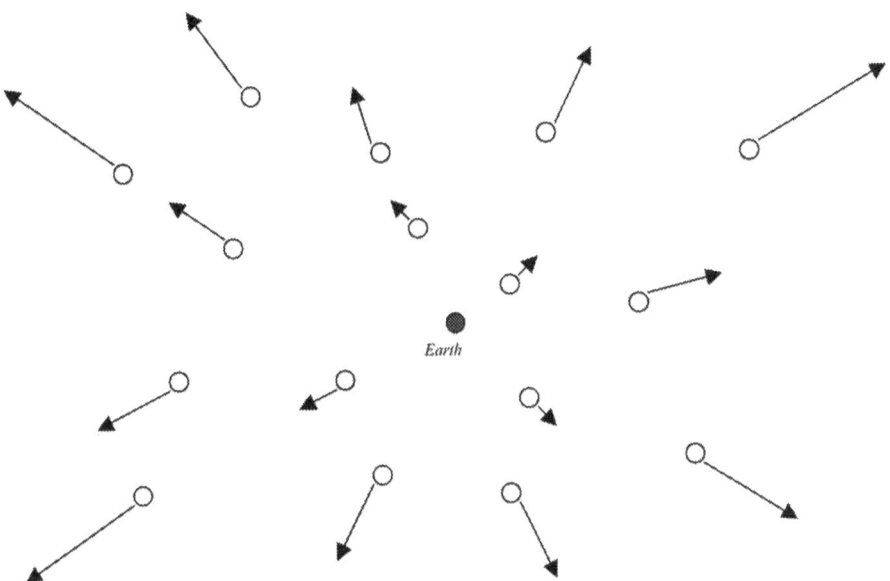

A representation of an expanding universe. The circles outside of earth represent galaxies at various distances. The arrows indicate the direction and magnitude of the velocities they have. Regardless of what direction we look from earth, everything is moving away from us at speeds that increase with distance.

As I mentioned briefly above, this expansion is actually turning out to be an *accelerating* expansion based on the most recent red-shift surveys. This has very large implications for the fate of the universe. For decades it has been debated whether the universe had enough mass in it to eventually slow down the expansion, and lead to future attractions of all matter that would eventually lead to a *big crunch* (the opposite of the big-bang). For a long time measurements seemed to indicate that the cosmos was very close to this *critical mass* (the exact mass to slow down and halt the expansion). However with our new accelerating expansion data, the universe would be what is called *open*. In an open universe, it is destined to expanded forever (and accelerate in our case) until every point in our cosmos would have a black void of a sky. Eventually all the nuclear fuel that is being fused in stars such as our sun would burn out and the stars would cool down and eventually stop emitting light as well.

The other piece of evidence for the big bang model is the observed cosmic background radiation. This is the radiation remnants of the initial Big Bang.

During the first instants of the big bang the temperature was an unimaginable 10^{32} Kelvin. At such an extreme temperature, matter as we know it did not exist. With time, the expansion from the big bang spread out this energy into larger and larger volumes of space, allowing the temperature (energy) to lower as there was more space-time to move around in (just as steam escaping from the enclosed volume of a tea-kettle quickly expands and cools down once it escapes). As this expansion and cooling continued, the creation of quarks took place, which are the building blocks of protons and neutrons (made up of three quarks each). Further expansion and cooling allowed protons and neutrons to bind together to form atomic nuclei. A key juncture was eventually reached at around 300,000 years after the big bang. At this time, the energy of the big bang had dissipated enough to allow the first photons of light to be emitted. Before this time, the cosmos consisted of ionized particles (atoms stripped of electrons) that did not allow the transmission of photons. In other words, the cosmos was not *transparent* to photons until 300,000 years after the big bang. Even after two hundred thousand years of expanding and cooling, the cosmos was still a scalding 100,000 degrees Kelvin. Prior to 300,000 years after the big bang, there was so much energy in the primordial soup of the cosmos as to not allow any electrons to fall into their probability shells around atoms. All of the atomic nuclei were stripped of their electrons. These free electrons in the early cosmos scattered any photons that were emitted, thereby making space *opaque* to light. Emitted photons were re-absorbed and did not travel any significant distances. Once the temperature of the cosmos dropped below 3000 Kelvin the first photons of what is now the cosmic background radiation were released, along with a snapshot of what the cosmos was like at that instant. Based on the estimated 13.7 billion year age of the universe derived from the expansion of space, we can predict the wavelength of these energetic photons from 300,000 years after the big bang should have stretched out to. As an object cools, the wavelength of its photons increase as the energy of a photon is inversely to its wavelength. Theory predicts that on their journey here, these first photons wavelength would have lengthened to the microwave part of the electromagnetic spectrum. That is exactly where we detect them, at microwave wavelengths, indicating they have cooled to around 2.7 degrees Kelvin on their 13.7 billion year journey to earth.

The cosmic background radiation is very uniform in all directions as we would expect from a rapid expansion that has no central point (remember from the discussion above that the big bang was an explosion that occurred everywhere). Cosmologists and astronomers like to call things that are evenly distributed like this *homogeneous.* To account for this high degree of uniformity, many cosmologists feel that there was an unthinkably rapid expansion in the earliest phases of the universe which spread out all of the irregularities from the big bang across huge distances—thereby greatly reducing their magnitude and giving rise to the homogeneity that we observe today. This period of rapid expansion in the cosmos is referred to as *inflation.* Theories that incorporate this concept are referred to as *inflationary theories.*

When people think of something "inflating" they typically think of balloons, or bicycle tires, not an entire universe. While it may take a couple of minutes to inflate a bicycle tire, or a few seconds to inflate a party balloon—the leading inflationary models of the cosmos have it expanding by factors of 10^{30} to 10^{100} within a *fraction of a second.* Despite its name, if a version of the inflationary big bang model turns out to be correct, as evidence is leaning towards, then it was truly an explosive expansion beyond anything that can be conceived by the human mind. So explosive was this initial expansion of the cosmos, that there are probably regions of the universe beyond what we are able to observe that will be forever hidden from our instruments. It is this enormous expansion that has made any given point in space appear so "flat". Einstein showed that space-time distorts in the presence of matter, if an incredibly rapid expansion of all the matter in the cosmos occurred early on, then it makes sense that everything seems so evenly distributed regardless of what direction we point our instruments in the sky.

Essentially, inflationary theories are cosmic theories that take into account the uncertainty principle of quantum physics. We will explore the concept of quantum physics and the uncertainty principle in more depth later. In a nut-shell, the uncertainty principle dictates that there always must be a minimal amount of irregularly or uncertainty in all aspects of nature. In the early instants of the big bang according to inflationary theories, these minute quantum irregularities were present, and were spread out over the entire visible universe by the inflation process to give rise to the small irregularities that we observe today in the cosmic background radiation's "fingerprint" of the early cosmos. A good explanation of this is presented in Michio Kaku's book Parallel Worlds, on page 101: "Since the inflationary theory is a quantum theory, it is based on the Heisenberg uncertainty principle, the cornerstone of quantum theory. (The uncertainty principle states that you cannot make measurements with infinite accuracy, such as measuring the velocity and position of an electron. No matter how sensitive your instruments are, there will always be uncertainty in your measurements. If you know an electron's velocity, you cannot know its precise location; if you know its location, you cannot know its velocity.) Applied to the original fireball that set off the big bang, it means that the original cosmic explosion could not have been infinitely "smooth." (If it had been perfectly smooth, then we would know precisely the trajectories of the subatomic particles emanating from the big bang, which violates that uncertainty principle.) The quantum theory allows us to compute the size of these ripples or fluctuations in the original fireball. If we then inflate these tiny quantum ripples, we can calculate the minimum number of ripples we should see on the microwave background 380,000 years after the big bang. (And if we expand these ripples to the present day, we should find the current distribution of galactic clusters...)".

The Creation and Evolution of the Cosmos—a Current Picture

Armed with our new knowledge of the big bang and inflation, we can now pull all the pieces together and provide an outline of the birth and evolution of the cosmos based on our current scientific understanding. Such a big bang model, taking into account the inflationary phase of the early cosmos is referred to as the *inflationary cosmological model.*

The cosmos in our current understanding started 13.7 billions years ago as a nugget of space-time of unimaginable density and temperature. During the first instants of the big bang at the first calculable instant of time, the *Planck time*, 10^{-43} seconds, the temperature was a searing 10^{32} Kelvin. Exploding from this searing nugget came the beginning of space and time as we know it. The big bang started the concept of what we refer to as time and what we refer to as space.

After the big bang, the hot plasma of constituent particles flowed outward in all directions, cooling in the process. The first tiny clumps of this superheated plasma started to form. At round 10^{-35} seconds, the inflationary mechanism initiated and expanded the cosmic soup exponentially by a factor of at least 10^{30}. At around 10^{-6} seconds (millionth of a second), this cooling of the plasma was enough to allow quarks to gather in sets of three, thereby instigating the formation of protons and neutrons. Not long after this time, at 10^{-4} seconds, further cooling allowed for the creation of the first atomic nuclei. This is known as *primordial nucleosynthesis.* From this era, the light elements of hydrogen, helium, lithium, and deuterium (sometimes referred to as "heavy" hydrogen) were created. This process continued for the next three minutes, until the temperatures dropped to a point where there was no longer enough energy for any more elemental creation.

From this stage, we enter an epoch of cooling and expansion that lasts for 300,000 years. Up to 300,000 years, there was too much energy (temperature) for any electrons to become captured by any of the newly created atomic nuclei. At 300,000 years, there was another dramatic change. Finally, the energy levels dropped to a point where electrons were able to be captured by atomic nuclei and the electromagnetic forces began to balance out. The messenger particle for the electromagnetic force, the photon, is now able to move freely through the cosmos without being annihilated or absorbed by the hot plasma of the big bang. This key moment is when the universe become transparent to photons, and what is now the cosmic background radiation was first released.

The epoch from 300,000 years to 13.7 billions years (the present) will have the first stars forming, and the first galaxies forming (and eventually clusters of galaxies) from those subtle instabilities and aggregations of matter that happened in the first instants of the big bang, and were rapidly expanded via inflation. These first rounds of stars were formed solely from the hydrogen, helium, lithium and deuterium that were present. As these stars went through their life cycles, they created successively heavier and heavier elements in their cores to fight off being collapsed by gravitational forces. Ultimately by means of supernovae explosions, the rest of the elements were forged and ejected into space to be introduced into the matter clouds of further generations of stars—eventually to become the building blocks for life and intelligent beings.

With the current limitations of general relativity and quantum theory, we are unable to probe any farther back than the Planck time. Superstring and M-theory hold hopes of providing more insight as to what may have occurred previous to this time. Some of these will be outlined in this book, however much of that research is still very nascent and the results are still inconclusive.

Olber's Paradox

This homogeneity of the cosmos used to bring up a strange paradox that was debated for many years prior to the discovery of the expansion and acceleration of the universe. This dilemma, called the *blazing sky effect* or *Olber's paradox* was actually a thought experiment. In this thought experiment, an infinite universe with homogeneous (even) distribution of stars and galaxies was imagined to lead to a night sky as bright as the surface of a sun. The reason for this strange conclusion is that if stars and galaxies containing stars were evenly distributed over an infinite volume, no matter what line of sight you picked to look at in the sky, no matter how small, that line of sight would eventually intersect the brilliant surface of a star. Interestingly, the first known correct solution to Olber's paradox was discovered by the writer Edgar Allan Poe (Parallel Worlds, pg 28):

"Were the succession of stars endless, then the background of the sky would present us an uniform luminosity, like that displayed by the Galaxy—since there could be absolutely no point, in all that background, at which would not exist a star. The only mode, therefore, in which, under such a state of affairs, we could comprehend the voids which our telescopes find in innumerable directions, would be by supposing that the distance of the invisible background is so immense that no ray from it has yet been able to reach us at all."

With the discovery of the cosmic expansion and acceleration, we have a handy solution to the blazing sky effect. Since the cosmos is expanding and accelerating, even if it is infinite, we are only able to see the events of 13.7 billions years worth of its evolution. The light from more distant objects past 13.7 billion light years has not had time to reach us yet (remember that Einstein found that light is the highest velocity possible), and keeps our sky nice and black instead of being as bright as a sun's surface.

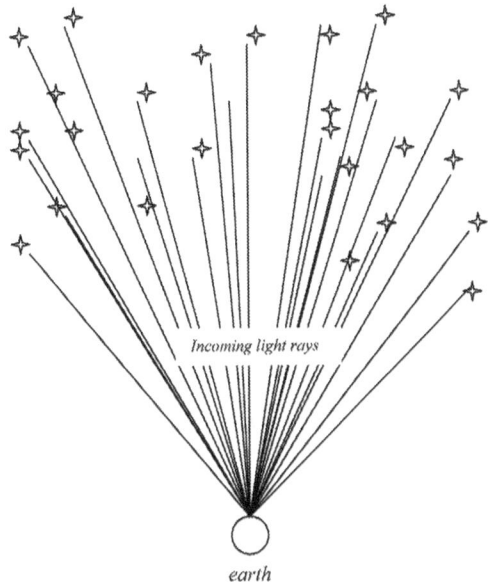

earth

Olber's paradox: if at any given point in space you intersect the surface of a star, the night sky should be as bright as the surface of a star. The key solution to this problem is that the speed of light is finite, not infinite, and that stars must have a life cycle of their own, where they stop producing light. So while it may be true that at any given point in space you would intersect with the surface of star if given enough time; those stars have finite lifecycles that they are producing starlight, so the light from stars that are burnt out will not reach earth, thereby solving Olber's paradox.

The WMAP satellite

The Wilkinson Microwave Anisotropy Probe or WMAP satellite is what has allowed Cosmologists to determine the age of the cosmos to unprecedented accuracy. With the Hubble space telescope techniques prior to WMAP's launch in 2003, the age of the cosmos was known to within 10%-20% accuracy. With the increased sensitivity of WMAP, we can pin down this age to around 1% accuracy, thereby giving us our present day value of 13.7 billion years.

3

Measuring Distances

For much of the history of astronomy and cosmology people thought that the universe beyond the solar system was static. Without the aid of precision celestial measurement technology, this is very easy to imagine. On the scale of a human lifetime, everything that we see in the skies above is essentially static. You can look at the constellation of Orion, the hunter, and can rest assured it will appear the same when you look up at it ten years later. Its position will change due to the rotation of the earth, and the earth's rotation around the sun...etc, but the *shape* of it will remain fixed. You will not look up one evening after a long day of work, take a deep breath, look up and say "Huh. Why is Orion's belt straight tonight?" While this is essentially correct for our short human lifetimes, over the course of much larger timelines, the stars that make up our constellations were positioned differently. And, as time marches on, eventually, the stars of our familiar constellations will drift apart—eventually warranting a new set of constellations to match the new patterns of stars in their new positions.

Radar

Practically everyone has heard of the term "radar". Technically it stands for radio detection and ranging. For accurately measuring distances at the scale of our immediate solar system, this is the preferred tool, and a good place to start as it is relatively familiar ground. When radio waves are transmitted at the speed of light towards an object in the solar system (say, the moon), upon reaching that object the radio waves will bounce off that object in all different directions. Some of those waves will be bounced back towards the direction of the earth, and a sensitive radio receiver. Since we know the speed of light (186,000 miles per second or 300,000 kilometers per second) we can calculate the distance to the object very accurately. There are limitations beyond our solar system for this technique, since a small fraction of the total radio waves will echo back towards earth, and the farther those echoes have to travel, the more their signal spreads out and weakens, eventually to a point beyond detection.

Parallax

The reason we can get away with our *approximations* that the stars do not move under day to day observations (and create accurate star charts that do not need updating every year) is because of the *distance* they are from earth. As I mentioned in Chapter one, the scales in astronomy are exponentially larger than anything we are used to dealing with in our everyday experience. The classic example of how distance relates to changes in position is to place one of your index fingers about one foot from the center of your eyes (in reality is does not matter where your finger is, as long as you can focus on it—the middle is just more comfortable). Now start to blink alternately with your left and then right eye and observe how much your finger shifts in reference to the fixed background. Now move your index finger out to your arms length and repeat the same procedure. You will notice that your finger shifts over a smaller distance relative to the background. This phenomenon is referred to as *parallax*. This same effect applies to all objects, including stars. If you can see the difference in parallax very apparently with utilizing just the length of your arm (let's say three feet), it is easy to see how stationary an object can appear when it is *light-years* away from the observer.

With the parallax technique you can achieve more accuracy, and hence measure the distance to farther away objects by increasing the distance that is referred to as the *baseline* (you need to increase the baseline to increase accuracy—because the farther away an object gets, the smaller the object's shift gets due to parallax). In the above example with your index finger shifting around relative to its background when you blink, the baseline would be the distance between your eyes. Since there is a mathematical relationship between the distance to an object and the amount of shift that occurs when you look at on object from different points along a baseline, you can calculate how far the object is.

With parallax measurements, the problem is that the objects are light-years away. In order to observe the shifts in positions in our stellar neighborhood to determine distances, astronomers can make observations of the same object from opposite sides of the earth. This increases the baseline to the diameter of the earth, and increases the distances you can accurately measure. The largest baseline that can be created by astronomers is to take advantage of the earth's orbit around the sun and make observations of an object you wish to measure the distance to at six month intervals, when the earth has traveled half-way around its orbit. This allows you to create a baseline that is the diameter of the earth's orbit around the sun. As you might imagine, using the whole orbit of the earth as a baseline greatly increases the distances that astronomers can measure.

Below is a visual representation of parallax based on measurements from six month separations in earth's orbit:

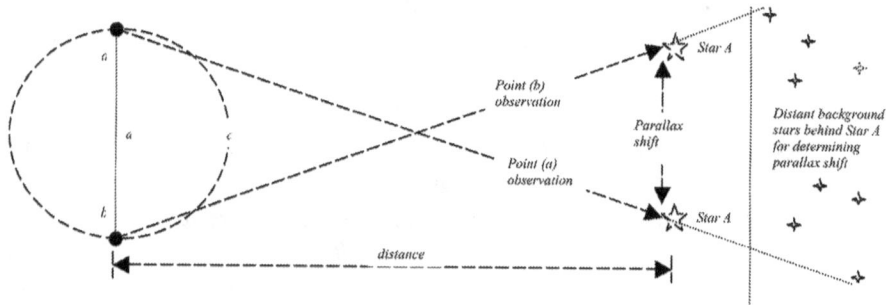

Example diagram of parallax based on measurements from the earths orbit at 6 month separations, where earth is at opposite ends of its orbit (*c*), at points (*a*) and (*b*). The great distance between points (*a*) and (*b*) in earth's orbit enables a large baseline (d) for which to measure the parallax shift of *Star A*. This amount of shift is determined from the more distance background stars behind *Star A* which appear fixed because of their much greater distance. The amount of parallax shift is proportional to the distance to the object, and the baseline (d) of the measurements.

Cepheid Variable Stars

Beyond the effects of parallax, there are other measurement tools that astronomers utilize to measure even greater distances. Certain types of stars vary their total light output in a very predictable way as variations in the fusion reactions in their cores fluctuate *periodically*. More specifically, with these types of stars the length of time it took them to brighten and dim was directly related to their brightness. The brighter the star was, the longer was its periodic cycles of dimming and brightening. These special stars are called *Cepheid* variables. If you take any light source, and graph out it's intensity over the course of time, you observe what astronomers call a *light curve*. If the intensity is unchanging with time, then you would have a straight line. However if the intensity is fluctuating as a Cepheid's light does, you would see rises and drops that start to form in your chart as time passes. If you kept measuring its light for long enough, you would eventually see that the graph starts repeating itself—that there is a pattern to light curve variations. Cepheid's repeating pattern in their light curves and their relation to their brightness was the key for astronomers to find the correlation between a Cepheid's light intensity as viewed from earth, and its distance.

Just as with any measurement tool, Cepheid's distance markers have their limitations. As anyone knows, the farther you get away from an object, the dimmer it appears—so at some distance you will not be able to see it. When you start getting out to the distances of galaxies, millions of light years and beyond, you need yet another tool: supernovas.

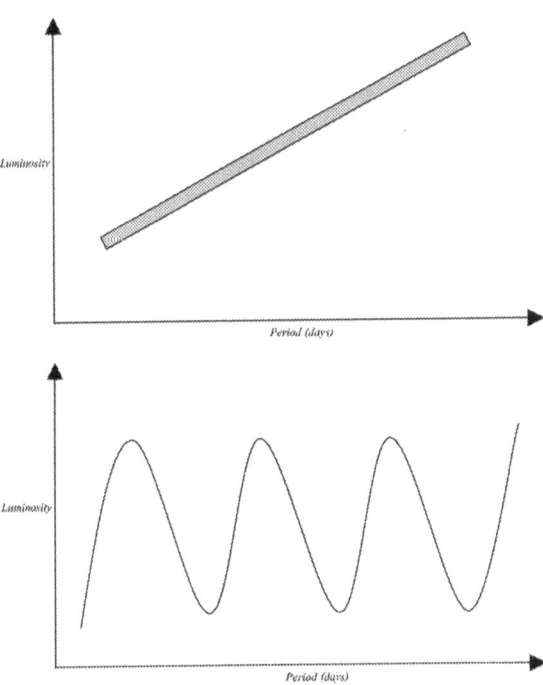

Cepheid variables as distance markers. The top chart shows an approximate representation of the correlation of the period of a Cepheid variable to its luminosity. The bottom chart is a sample of how the luminosity of a Cepheid fluctuates over time in a predictable manner. These characteristics of Cepheid's enable astronomers to determine their distance. If you find the period, this will give the luminosity. Once you measure the apparent brightness of the star, you can then determine the distance from the brightness and luminosity.

Supernovae

A supernova represents one of the most violent acts in the cosmos—the death of a massive star. While there are difference types of supernovas, they all boil down to the same basic scenario—a massive star that has run out of nuclear fuel to fuse together to hold off the forces of gravity that are trying to collapse it. The result is a huge explosion that produces enough light to equal the output of an *entire galaxy* for a brief time, before fading away. Again, the key here is linked to a supernova's light curve. Just as Cepheids have the same light curves, certain types of supernovas have the same light curves over time as well while they spill out their dying remnants of light towards us. Also, as with Cepheids, a correlation was found between how certain types of supernova's light decreased with intensity through time depending on their distance from us. If a supernova can produce an entire galaxy's light output concentrated into a single point then it is not hard to imagine that such a beacon could be seen for tremendous distances. Indeed, supernovas provide the one farthest *reliable* measuring stick in the astronomer's toolbox—allowing measurements out to galaxies hundreds of millions of light-years away.

Red Shifts

But there are many objects in the cosmos that are spread out to a distance of several hundred million light years, and billions of light years (remember from chapter one that the universe is around 13.7 billions years old). To measure distances out to these great depths astronomers need to use a combination of all the previous tools described, and a new tool called *red shifts*. The term red shift has been around for long enough now to have trickled more into the mainstream knowledge base. The principle of a red shift is very similar, but not to be confused with, everyday phenomena that we *hear* all the time (instead of "see").

Most people know that the horn of a car that is speeding towards you while you are standing on a street corner sounds higher in pitch than when it goes by you and becomes lower in pitch. Another common experience is to hear a train whistle at a railroad crossing as the horn gets blown to warn of its presence. The same thing occurs—the train horn is higher in pitch as it is approaches your stationary car, and is lower in pitch once it passes your car and is heading away from you. With sound waves this effect is called the Doppler Effect or a Doppler *shift*. What is shifting is the *wavelength* of the sound waves as they get compressed in front of a train or car moving towards you, and stretched out behind them as they race away.

This is the same thing that happens with light. Light particles travel up and down through space much like sound waves from a car or train horn. There is often some confusion in the public eye on light behavior as it has properties of both waves and particles—so let us clear this up first before continuing on with the explanations of red shifts. Quite simply, light is a stream of particles—photons—not a wave. Often people will refer to wave-particle duality of light, and this can be misleading. Light is indeed a stream of particles; however if you are measuring for wave-like properties, it will behave light a wave (e.g. testing for interference), and if you are measuring it as a particle (e.g. individual photons impacting a phosphor screen for recording), then it will appear as a particle. *The act of measurement is what determines if the light will appear wave-light or particle-like.* I realize that this is a counter-intuitive statement, and it is a salient point behind the mystery of quantum physics, which we will be exploring in more detail later. When scientists are studying and explaining light behavior on scales that we see in everyday events (large scales compared to the tiny size of individual photons) the overall behavior of light can be simpler to understand as a wave since at those scales you are dealing with millions of photons at a time that can be *approximated* accurately by using wave descriptions. A good way to think about a photon of light traveling through space is to think of dropping a small ball into a tilted trough that is wavy. The ball (photon) has no choice but to stay in the wavy trough, but if you track out the balls location over time on a chart, it would trace out the wavy curves of the trough. At a fine enough scale, we would see that this is not a typical wave—it is a probability wave, which we will explore further. This probability wave is the result of our modern understanding of the behavior of photons as presented in quantum physics.

Another property of light emitted by objects is that astronomers can smear out their light with a prism or diffraction grating to study it in detail. A good example of a diffraction grating is to think of a compact disc and how light is scattered into a rainbow when light hits it from all the tiny parallel grooves that contain the information on the disc—a diffraction grating utilizes the same concept. The only difference is that a diffraction grating uses parallel straight lines to diffract the light instead of the curved parallel lines on a compact disc. This smeared out light from an object is called a *spectrum.*

The frequency of these wiggles is what corresponds to a different color. The lower the frequency (less wiggling) of light, the less energy it contains, and as the frequency increases (more wiggles), the energy it contains increases as well. So in our visible part of the spectrum, red has the lowest frequency and hence the least energy, and blue/violet light has the highest frequency and hence the most energy of the *visible* spectrum. There is a much larger range of frequencies that reside outside of the ones that our eyes are sensitive to. In order of lowest to highest frequency (lowest to highest energy) they are radio waves, microwaves, infrared, *visible*, ultraviolet, x-rays, and gamma rays. It is important to clarify that despite the wildly different names for each range of frequencies (wiggling), they are all descriptions of photons wiggling at different rates—that's all. The highest energy gamma rays are made of the same type of photons as the lowest energy radio waves. The entire range of wiggling (frequencies) that is available for a photon is called the *electromagnetic spectrum.*

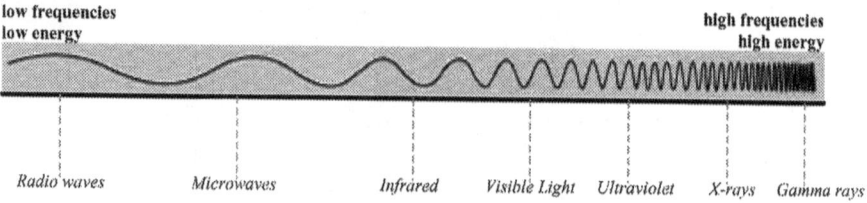

When astronomers smear out the light from objects they are observing, they see complex patterns of bright and dark lines that are called *emission lines* (the bright lines) and *absorption lines* (the dark lines). This occurs because all matter is made of atoms that emit and absorb photons of specific frequencies. These emission and absorption lines are unique for each element and are the equivalent of a fingerprint for that element. Combined, emission and absorption lines can be called *spectral lines.*

Once astronomers had mastered the techniques for acquiring and examining spectra (spectrum—singular form) for objects, something peculiar started coming out of these spectra when more and more distant galaxies spectra were studied. The familiar bands in the spectra that they were used to seeing for specific types of more common elements such as hydrogen, helium, and oxygen were not showing up where they were used to seeing them on the spectrum of those elements as recorded in a laboratory setting. When an object is made of many different elements, the most abundant elements in that object produce the most intense emission and absorption bands and are thus more easily detectable over great distances. Instead they were being *shifted* towards the *red* part of the spectrum while maintaining the exact emission and absorption band patterns. What was found is that light, just as our train and car horn examples earlier, gets "stretched" if it is traveling to us from a source that is moving *away* from us (or "compressed" when moving towards us).

The illustration below summarizes the concept of a spectral redshift:

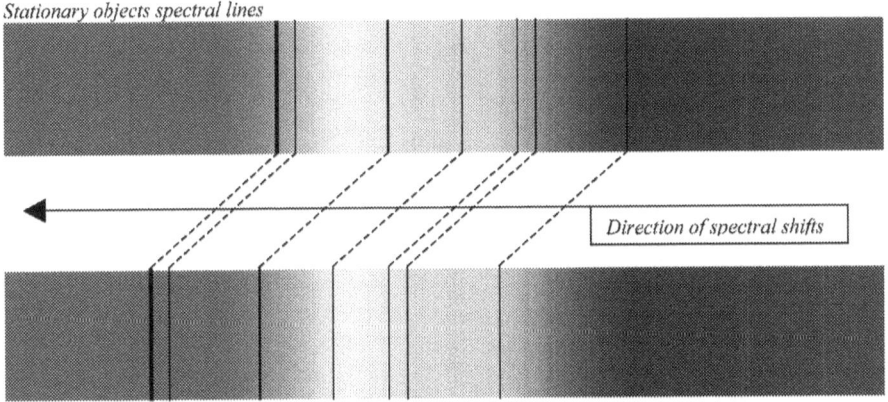

*Spectral lines measured as an object whose radial velocity is away from the observer. Note the shift in spectral lines compared to the stationary spectral lines. (*Note that if the radial velocity of the object was moving towards the observer, the lines would be shifted towards the blue/violet end of the spectrum). The amount of shift in the spectral lines is correlated to the radial velocity towards or away the observer.*

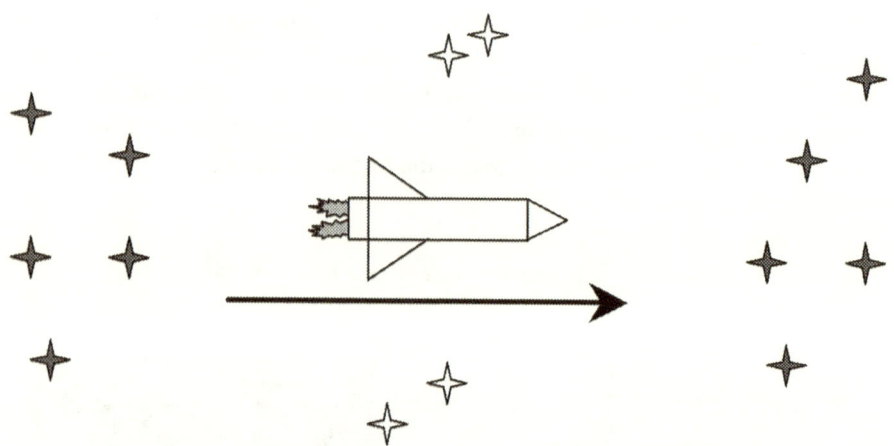

In a rocket that is speeding along in space at near light speeds, the stars in front of the space craft would appear to be more blue since the rocket ship is speeding towards the incoming radiation. This makes the ship move closer during each wave cycle of the incoming radiation, thereby shortening its wavelengths towards the blue end of the spectrum. Conversely, a view out the back of the speeding rocking will show the stars appearing more red as the rocket speeds away from the radiation during each wave cycle and thereby increases the wavelength of the light towards the red end of the spectrum. Any stars visible perpendicular to the direction of motion will appear unchanged, as the direction of rocket ship is not towards or away their incoming radiation, leaving it unchanged.

Once many galaxies' spectra started being observed and techniques for measuring spectral lines became more precise, it was found that, in general, all galaxies, regardless of what direction you were looking, had their spectral lines shifted significantly towards the red end of the spectrum. These red shifts meant that all of these galaxies were moving away from us at high speeds. Additionally, it was found that the more distant a galaxy you were observing the spectrum for, the *larger* these red shifts would become. So this meant that the universe was expanding faster when it was younger. The reason we can state this is due to the fact that the highest possible speed in the universe is the speed of light. Nothing can travel faster than the speed of light due to the mass increase with velocity that Einstein discovered. This means that if we have a handy time machine present for astronomy and cosmology. If the speed of light is the speed limit and photons are traveling at this speed, then the distance to any object in light-years is how far back in the past we are seeing the light emitted by that object. If we are looking at a galaxy that is 100 million light-years away, then we are seeing how that part of the universe appeared 100 million years ago.

Remembering that all of these red-shifts were occurring regardless of what direction astronomers pointed their telescopes, and that these red shifts had an increasing magnitude with distance, we can deduce that the universe is expanding away from us in all directions, and therefore started from a much denser and smaller amount of space than we see today. This is one of the key pieces of evidence for the big bang—the observed universal expansion of the cosmos. For some galaxies that are in clusters such as our Local Group—we sometimes see *blue shifts* occurring from the galaxies moving towards us due to the gravitational attractions to the other cluster members that overcome the effects of the larger universal expansion. In fact, the Milky Way's nearest neighbor in the Local Group, the Andromeda galaxy, is headed for a collision with the Milky Way in the distant future.

In recent years, the techniques and technologies behind detecting the shifts in spectral lines towards the red or blue has become sensitive enough so as to allow the detection of larger mass (Jupiter type—but many times more massive) planets orbiting other suns in the Milky Way. As a large planet orbits its parent sun, it "tugs" that star towards us and away from us just a little bit, depending on where the larger planet is in its orbit (think of an ice skating couple with their arms locked together—spinning on the ice—as they spin, their center of gravity "wobbles" back and forth because of the weight difference between them). Astronomers can now detect these little changes in velocity (wobbles) and, based on data for that sun's mass and distance, determine the mass of the planet that is causing the observed "tugs" (as of this writing this technique has identified over one hundred large mass planets around other stars, providing evidence that planet formation is much more common than previously expected).

Superluminal (faster than light) Velocities and Redshifts Revealed

There is in fact *superluminal*, or faster than light velocities that are routinely observed in the cosmos. How is this possible when Einstein's relativity specifically sets the cosmic speed limit at *c*, the speed of light? The clearest explanation I have recently found was in the <u>March 2005 Scientific American (pages 40-42)</u>: "Astronomers have observed about 1,000...objects [galaxies] receding from us faster than the speed of light. Equivalently, we are receding from those galaxies faster than the speed of light. The radiation of the cosmic microwave background has traveled even farther…When the hot plasma of the early universe emitted the radiation [cosmic background radiation] we now see, it was receding from our location at about 50 times the speed of light….The idea of seeing faster-than-light galaxies may sound mystical, but it is made possible by changes in the expansion rate. Imagine a light beam that is farther than the *Hubble distance* [the *Hubble distance* is the distance at which an object is receding from us at the speed of light due to the expansion of space-time] of 13.7 billion light-years and (is) trying to travel in our direction. It is moving towards us at the speed of light with respect to its local space, but its local space is receding from us faster than the speed of light. Although the light beam is traveling toward us at the maximum speed possible, it cannot keep up with the stretching of space. It is a bit like a child trying to run the wrong way on a moving sidewalk. Photons at the Hubble distance are…running as fast as they can to stay in the same place. One might conclude that the light beyond the Hubble distance would never reach us and that its source would be forever undetectable. But the Hubble distance is not fixed, because the Hubble constant on which it depends, changes with time. In particular, the constant is proportional to the rate of increase in the distance between two galaxies, divided by that distance…In models of the universe that fit the observational data, the denominator increases faster than the numerator, so the Hubble constant decreases. In this way, the Hubble distance gets larger. As it does, light that was initially just outside the Hubble distance and receding from us can come within the Hubble distance. The photons find themselves in a region of space that is receding slower than the speed of light. Thereafter they can approach us…Special relativity applies only to "normal" velocities—motion *through* space [as opposed to the global expansion *of space itself*]. The velocity in Hubble's Law is a recession velocity caused by the expansion of space, not a motion *through* space….Having a recession velocity greater than the speed of light does not violate special relativity. It is still true that nothing ever overtakes a light beam."

4

Based on the incredible magnitudes of volume occupied by our observable cosmos, and the fact that is has been evolving for 13.7 billion years, it is no surprise that there is a myriad of different types of objects contained within it. This chapter will summarize the different types of extra solar (outside of our solar system) objects that astronomers study. I will present objects based on their proximity to earth starting from the closest objects housed within our galaxy (and all galaxies in the cosmos) and ending with the farthest objects we are currently able to observe and classify.

Stars (general overview)

The first type of object on our journey is a star. Technically stars are massive aggregations of dense gas (mostly hydrogen and helium) whose associated gravitational field is strong enough to compress the gas to a point where nuclear fusion of lighter atoms into heavier atoms takes place. The energy released from nuclear fusion is copious (as we unfortunately know here on earth from the advent of hydrogen bombs) and exerts an outward pressure between the atoms of the gas cloud that balance out the inward force of gravity, thereby providing a balance of forces for as long as the fusion process can be sustained. As you may remember from your high school physics class, if you compress a volume of gas, it heats up due to the atoms increased tendency to bump into each other in the compressed volume and release that energy in the form of heat. This same basic principle is what triggers a star to form. If you have a large enough mass of gas that starts contracting under its down gravity (sometimes this contraction of the gas gets triggered from an external event such as a nova or supernova shockwave (which will be explained later in this chapter) it will become dense enough for the atomic nuclei to fuse together. The initiation of this nuclear fusion in the core of an ultra-dense mass of gas is where the definition of a star begins: a sustained internal fusion reaction. The reason that stars are so huge is simple—gravity is weak as compared to the other forces in nature—being less than a trillionth of a trillionth of the nuclear Strong force which binds the protons and neutrons in the atomic nuclei together. Due to this weakness, it takes an enormous amount of mass to generate enough of a gravitational force to hold together a star when the fusion process is constantly trying to expand it outwards.

From this fusion reaction, a tremendous amount of light, heat, and outward pressure are released in order to hold off further gravitational collapse. As we all know from humankind's development of the atomic fission and fusion (hydrogen) bombs, the energy released from these types of nuclear reactions are enormous. To get a grasp on a star's energy output from fusion consider this: even in a low-mass star, the energy output from its fusion reactions in the core is powerful enough to *exceed* the energy output of the detonation the entire world's nuclear arsenals simultaneously for every *second* that this star shines (note that even the briefest of stellar life-cycles persist for millions of years, with most going into billions {thousands of millions} and even trillions {thousands of billions} of years). In the end however, for all these stars the eternal constant of gravity ultimately wins. At some point, be it millions, billions, or trillions of years, a star will run out of atoms to fuse in its core and gravity will collapse the core and extinguish the nuclear fires.

Stars cover a large range of sizes and temperatures outside of what we experience in our solar system with the sun. There are seven *spectral classes* that stars can fall into based on their surface temperature. These spectral classes, according to highest to lowest surface temperature order, are: **O, B, A, F, G, K, M** (I have bumped into a couple of books that refer to the following phrase *"(O)h (B)e (A) (F)ine (G)irl [or (G)uy] and (K)iss (M)e"* to remember them). Within these classes are 10 subsets (0 to 9) that specify the relative surface temperature within a particular spectral class (e.g. G0, or A3, or M9…etc). The lower this sub-number is, the higher the surface temperature of the star within that spectral class. For example, an A3 star has a higher surface temperature than an A4 star, but a lower surface temperature than an A2 star…etc.

In this classification system, our sun is a spectral class G2 star, with a surface temperature of around 6000 *Kelvin* or 6000 K. The Kelvin scale is another temperature measurement scale like the more familiar Fahrenheit or Celsius systems. Kelvins are simply the Celsius temperature plus 273.15. The lowest possible temperature *theoretically* possible is 0 Kelvin, or-273.15 Celsius. This temperature is called *absolute zero*. At this temperature *all* atomic motions (what temperature is really measuring—the energy in atomic motions) are zero. While this is the lowest theoretical temperature possible, in reality it is impossible to stop all atomic motions and achieve true absolute zero due to the uncertainty principle of quantum physics. Stars range in surface temperature from around 3000 Kelvin up to around 30,000 Kelvin. In their cores, where the nuclear fusion is fusing lighter elements into heavier elements (Hydrogen into Helium is the most common in stellar cores) the temperatures are much higher. The simplest fusion of hydrogen atoms to yield helium atoms requires temperatures of at least 100 million Kelvin (100,000,000 Kelvin).

As there is a large range of temperatures for a star, there is also a large range of masses as well. All stars are truly massive objects by human standards. Our sun, a fairly *average* star has a mass of around 1.99×10^{30} kilograms (where 1 pound equals 0.4536 kilograms). The sun's *luminosity* or the measure of how much energy leaves its surface *every second* is 3.9×10^{26} Watts. And these numbers are for the *average* example of our sun. Stars range in mass from around one-tenth the mass of our sun to over 20 times the mass of our sun. For reference our earth has a mass of 5.97×10^{24} kilograms, which is about 333,333 times less massive than the sun. There is evidence that in the early cosmos when the first round of stars were ignited, that they were truly massive, being upwards of *hundreds* of times the mass of our sun. Larger stars that are 5 to 50 times the *diameter* of our sun are referred to as *giants*. Beyond this, there are *supergiants* that have diameters between 50 and 500 times the diameter of our sun (an object that has a diameter 500 times that of our sun would be 6.96×10^{13} km.

The more massive a star is, the more elements it is able to synthesize in its interior. The interior of a massive star can be thought of as an onion, with each layer representing another type of element that is being fused. Despite the tremendous pressures and temperatures that are able to be generated in stellar interiors, there is a limit to this fusion of heavier elements. Once iron is being fused, stars of any size are not able to generate enough pressure and energy to fuse any elements that are heavier. For this, a massive star must suddenly collapse and rebound explosively as a supernova. A supernova then has the energy needed to fused elements beyond iron. The illustration below outlines this.

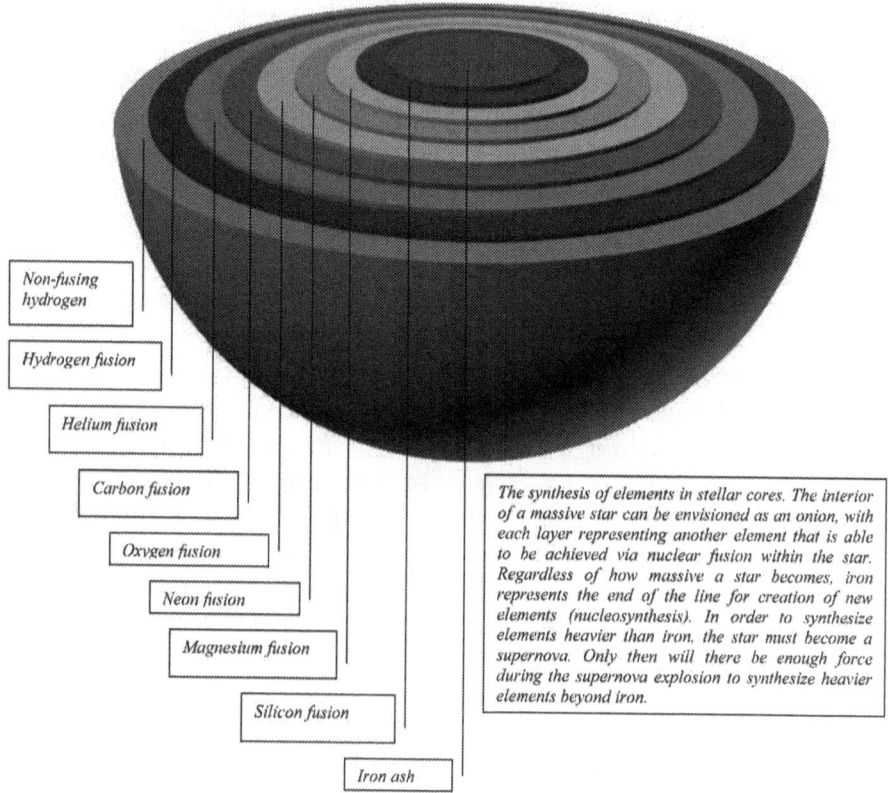

Non-fusing hydrogen

Hydrogen fusion

Helium fusion

Carbon fusion

Oxygen fusion

Neon fusion

Magnesium fusion

Silicon fusion

Iron ash

The synthesis of elements in stellar cores. The interior of a massive star can be envisioned as an onion, with each layer representing another element that is able to be achieved via nuclear fusion within the star. Regardless of how massive a star becomes, iron represents the end of the line for creation of new elements (nucleosynthesis). In order to synthesize elements heavier than iron, the star must become a supernova. Only then will there be enough force during the supernova explosion to synthesize heavier elements beyond iron.

Dwarf Stars

As mentioned above, the sun is an average star. To astronomers, stars that are similar in size or smaller than our sun are called *dwarf stars*. In this dwarf star category, there are sub types as well. A *white dwarf* is the end result of a low-mass star such as our sun exhausting its nuclear fuel. When the nuclear fusion reactions cease, and the core of a low-mass star is unable to hold off the gravitational forces any longer. The core collapses down into an incredibly dense object that is about the size of our planet earth (possible slightly smaller). As you may remember from earlier in the book, the sun can contain about 1 million earths. Imagine having all of the sun's mass (actually the sun loses some mass in a series of helium "flashes" that contribute towards the generation of a planetary nebula) compressed down into a volume that is roughly one *million* times smaller. This resulting earth-sized object is the burn-out remnants of a stellar core such as our sun, and glows bright white due the immense surface temperatures (for a star the size of our sun, the resulting white dwarf would have a surface temperature on the order of 50,000 Kelvin). This high surface temperature translates into a white light, and is what gives *white* dwarfs their name. White dwarfs no longer have the capability of producing their own internal fusion reactions. They are more akin to the glowing embers that remain after a fire is burnt out, that are slowly cooling down as they radiate their energy away into their surrounding space. The resulting densities in a white dwarf are truly immense. A *marble-sized* piece of a white dwarf on earth would weight considerably more than an entire car. The repulsive forces between the electrons in the white dwarf's atoms are what end up halting the gravitational collapse. This repulsive force is referred to as the *electron degeneration pressure*.

Another type of dwarf star is a *black dwarf*. A black dwarf is simply the end product of a white dwarf slowly radiating away its latent energy. Once the excess energy from its nuclear fusion days has been radiated, it cools, and becomes dimmer and dimmer until it becomes essentially *black*.

The third type of dwarf star is referred to as a *brown dwarf*. In this category of dwarf stars, instead of seeing the aftermath of a star after its nuclear reactions have ceased, we are looking at objects that never acquired the mass (around one-tenth the mass of the sun) and internal core temperatures required to start up their nuclear fusion reactions. A brown dwarf is a failed star. Just as a white dwarf will eventually turn into a black dwarf after cooling down, the same will happen to a brown dwarf. The heat that it did manage to generate from gravitational compression will slowly radiate away into space and cool it down into a non-visible object as well (however with a fraction of the density of the progenitor white dwarf that was once the size of our sun).

Neutron Stars

For stars that are massive enough to trigger supernovae explosions that do not completely destroy the central core (around three to five solar masses), it is possible for an ultra-dense, exotic object called a *neutron star* to be left behind. Despite their exotic properties, they are the result of the same gravitational collapse that leads to our white dwarfs described earlier. The main differences between a neutron star and a white dwarf are that in a neutron star the mass of the parent star from which it was attached to earlier in its evolution was significantly larger than that of our sun, and secondly, due to this increased mass, the resulting compression in a neutron star is much higher than in a white dwarf. The name neutron star arises from the fact that electrons are compressed into protons with such force during the gravitational core collapse as to combine them together into neutrons (electron have a negative-1 charge, and a proton has a +1 charge so that when they are combined under tremendous pressures and temperatures they form a neutral neutrons). After this compression is done, all that remains of the parent star is a core of super-compressed neutrons—or a *neutron star.*

Once again, the magnitude of this compression in neutron stars is worth elaboration. With a white dwarf, you have an earth-size object that contains roughly the mass of the sun. In a neutron star, you have an object that contains a mass significantly larger than that of the sun, in an object that is around *20 kilometers* across. If somehow you could stand on the surface of a neutron star you would be instantly flattened thinner than the thickness of a *human hair* due to its immense gravity. To get a better perspective on the density of neutron stars in earth-terms, I will quote a section from the late Carl Sagan's classic book Cosmos (page 239) (please note that I have added the *italics* for my emphasis): "Neutron star matter weighs about the same as an ordinary *mountain* per *teaspoonful*—so much that if you had a piece of it and let go (you could hardly do otherwise), it might pass effortlessly *through the Earth* like a falling stone through air, carving a hole for itself completely through our planet and emerging out the other side…If a piece of neutron star matter were dropped from nearby space, with the Earth rotating beneath it as it fell, it would plunge *repeatedly* through the rotating Earth, punching *hundreds of thousands* of holes before friction with the interior core of our planet stopped the motion." This is a mind boggling density approaching a *billion* times the density of a white dwarf. A *pea* sized piece of neutron star matter would weigh *millions* of pounds on earth."

In addition to this incredible density, neutron stars have another characteristic—a strong magnetic field. Just as our earth, with its iron rich core, generates a magnetic field that helps protects us from harmful cosmic rays and solar flares (and gives us our aurora borealis or "northern lights"), stars are also home to large magnetic fields that are many orders of magnitude greater than earth's. Just as the density increases as you compress something into a smaller volume, so does its accompanying magnetic field. Remember back to high school science class when you took a magnet, placed a piece of paper on top of it, and sprinkled iron flakes onto the paper. The iron flakes pattern themselves on the paper in relation to the *field lines* of the magnet underneath. If you imagine compressing that magnet into a tiny size, the density of those field lines would become greater as you compressed the magnet down in size. Put another way, as you compress a magnet you are increasing the number of field lines per unit area. This is the same concept that is in effect for neutron stars. The same magnetic field of their "parent" stars has been compressed down into an object that is around 20 kilometers across. This corresponds to increasing the magnetic intensity exponentially. Neutron stars have magnetic fields that are on the order of a *trillion* times that of earth's magnetic field.

Pulsars

A *pulsar* is quite simply a rotating neutron star. The earth rotates about its axis every 24 hours. Our sun and all stars as well, have their own rotation rates (usually weeks in length for average stars). Since the "parent" star that formed the neutron star would have a rotation that was weeks in length, as it fusion cycle shut-down and the core-collapse initiated, the rotation rate *increases* as the radius of the object *decreases*. The classic example of this is a figure skater who spins faster and faster as they draw their arms in towards their center of mass. On the scale of a star two times the mass of the sun getting compressed down into a neutron star that is about 20 kilometers across, that corresponds to rotation rates increasing from weeks to *fractions of a second.* This is what we see in pulsars—neutron stars that are spinning several times a second.

Coupled with this rapid spinning are the powerful magnetic fields that are a *trillion* times stronger than earth's. These magnetic field lines are dragged along and spin with the neutron star as well. As with those same magnets that you played around with in science class that had north and south "poles", the same applies to earth, sun, and neutron star magnetic fields. High energy charged particles around the neutron star get tangled up in the magnetic field lines, and break free at the "poles" of the magnetic field where the field lines converge. At the poles of neutron stars, these charged particles are expelled from the surface at high speeds out into the surrounding space. This creates a sort of spinning two-sided "light house" of sorts. If one of these "beams" of high energy particles happens to cross the plane of our solar system, then astronomers are able to detect them as rapid "pulses" through their telescopes. Hundreds of pulsars have been detected over the years, and the photons of these pulsar's "beams" have been discovered at all the wavelengths of the electromagnetic spectrum, from radio and visible waves, all the way up to high energy gamma rays (despite some pulsars emitting in the visible part of the electromagnetic spectrum, we are unable to see these pulses without special equipment since they pulse too rapidly for our eyes to detect them [on the order of several times per second]).

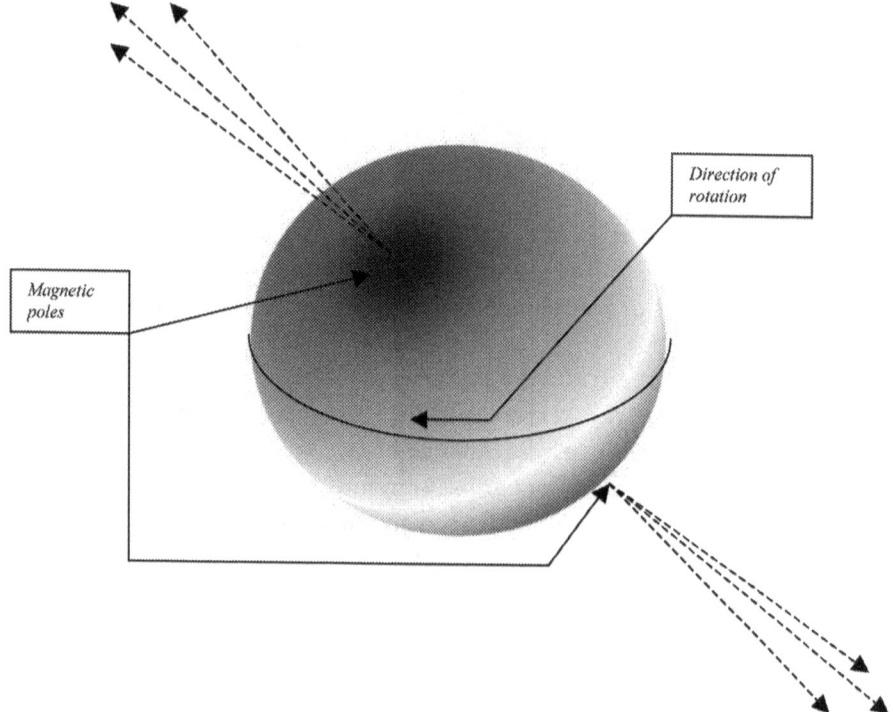

The lighthouse model of a pulsar. As the parent star contracts and explodes as a supernovae, the remnant will be a neutron star if the parent star lacks the mass to make a black hole. Just as an ice skater will spin faster as they pull their arms in towards their body, the parent star's rotational velocity will increase as it shrinks in diameter, giving rise to the extremely rapid rotational rates associated with pulsars. In addition to rapid rotation, neutron stars will have powerful magnetic fields that allow charged particles to shoot away from the poles of their magnetic fields as they rotate. When these rapidly spinning jets of energy happen to point at earth during the course of rotation then that is what we detect on earth as the "pulse" of a pulsar. The only difference between a neutron star and a pulsar is that the jets of a pulsar intersect with earth where they can be detected.

It should also be noted that as a pulsar keeps emitting charged particles via these "beams" from its magnetic poles, it is losing mass. This loss of mass must be accompanied by a slight decrease in its rotation period due to the conservation of momentum. It is akin to our spinning figure skater from earlier letting their hands splay out just a little bit, and hence slowing their rotation accordingly. Despite this loss of mass and decrease in rotational velocity with time, stable pulsars remain the most precise clocks in the universe, beating out the best atomic clocks on earth.

Any reader interested in getting involved in pulsar research and the detection of gravity waves may download Einstein@Home software for free from Berkley at http://einstein.phys.uwm.edu. It utilizes your home computer's idle time to scan data from spinning neutron stars (pulsars) for evidence of *gravity waves* as predicted by Einstein's theory of relativity. A gravity wave is a "ripple" in the fabric of space-time that travels at the speed of light. Einstein predicted that such waves should be generated whenever any object with mass is accelerated. These waves are extremely subtle unless a sufficiently massive object is accelerating, and only now have measurement technologies become sensitive enough to start testing for their existence. This is a very convenient, unobtrusive way for anyone with a computer and an internet connection to provide important contributions to cutting edge scientific research.

Magnetars

An interesting sub-category of neutron stars are called *magnetars*. These recently discovered objects have the strongest magnetic fields that have ever been measured (hence the name *magnet-ars*). Magnetars have extremely rapid rotation rates. Due to the density of the magnetic field lines and the rapid spinning, the lines get "twisted" together in the central regions of the magnetar. These twisted field lines eventually get so strong (around 1000 times stronger than the magnetic fields of pulsars) and packed together that they literally break apart the crust of the star's surface, suddenly releasing massive amounts of energy that can be detected by astronomers on earth as a "burst" of gamma rays. Even at tens of thousands of light years distance, magnetars can produce substantial effects on the earth's outer atmosphere as the high-energy gamma rays slam into it. Pound for pound, the gamma ray bursts released from the energy pent up in their magnetars magnetic fields are among the most powerful objects in the cosmos.

Earth's Magnetic Field

It is interesting to note that Earth's magnetic field is quite dynamic as well despite its relatively small magnitude as compared to a magnetar. A great article on this variability and the source for the following summary can be found in <u>Scientific American—April 2005</u> (Pages 51-57) that the interested reader will profit from. Earth's magnetic field is produced from the convection and turbulent flows in the molten iron contained in the Earth's outer core circulating around a solid iron inner core. This convection arises from the temperature differences between the parts of the outer core that are next to the 5000 Kelvin inner core (latent heat left over from the earth's initial formation) and the outskirts of the outer core (next to the mantle) material which are cooler since they are a farther distance from the inner core's heat. As we know in our everyday experiences, hot air rises, and cold air sinks. The same rules apply to liquids and to molten iron in the outer core as well. The molten iron next to the inner core gains heat energy and hence buoyancy, which makes it slowly rise outwards towards the mantle. During this journey outwards, the molten iron loses heat energy as its distance from the hot inner core increases. Eventually, it will cool to a point that it becomes denser, and less buoyant, than its surrounding outer core material, and start sinking back down towards the inner core again…etcetera. This convective process keeps the outer core in constant motion and, with the earth's rotational forces (more specifically the *Coriolis effect*) keeps the earth's magnetic field intact. The laws of electromagnetism dictate that any motion that occurs within an electrically conducting fluid will have an associated magnetic (and electric) field. And just as our iron flakes from earlier converged at the poles of our bar magnet, where the lines converge in the greatest numbers on their way into the earth's core are what we associate with our magnetic north and south poles.

What is largely unknown is that the "north" and "south" of our geomagnetic field has reversed several times in the past. Magnetic minerals that go from molten to solid states (passing through their *Curie point*) upon cooling will line up their molecules just like a series of tiny compass needles. Once such a magnetic material has cooled into a solid, whose tiny compass orientations are locked in time and can provide clues as to when magnetic reversals occurred. Over the earth's 4.5 billion year history, this type of geomagnetic reversal is believed to have occurred *hundreds of times*. While the computer models of these reversals are currently being studied, they are restricted to studying the effects of magnetic fields based on simple convective flows. Computers do not yet have the speed to three-dimensionally reconstruct turbulent flows within the outer core and the predicted effects on the earth's magnetic fields (as of this printing, models of this type of turbulent flows were restricted to 2 dimensional models, which are less computer intensive). The details of how these reversals take place are still in their early stages. The most recent models suggest that the earth's rotational effects on the fluid flow in the outer core, the Coriolis effect (this deflects what would normally be straight line motions in a non-rotating body into curved or deflected paths induced by the rotation of the body) cause deflections and twisting in the convective cycles, which also twists and bunches up their associated magnetic field lines. Eventually, these bundles of twisted (and reversed) field lines gain enough strength to make their way to the surface of the earth and flip the whole geomagnetic field. Models indicate that this process can occur in intervals as short as 9000 years.

Black Holes

Perhaps no other exotic object in astronomy is more profound than a *black hole*. In order to correctly explain these bizarre objects we must first remember two main points of Albert Einstein's Theory of General Relativity. First off, Einstein found that *everything* is subject to the force of gravity, even *light*. Secondly, he found that the highest velocity obtainable for any object is the speed of light. Armed with this data we can begin to explore the concept of black holes.

A star similar in mass to our sun will eventually burn out (billions of years from now), and compress down into an earth sized white dwarf. A star two times the mass of the sun will burn out and compress down into a 20 kilometer neutron star. A star even more massive, say four times the mass of our sun, will compress down into an object so dense, and be accompanied by such intense gravity, that even *light itself cannot escape*. This object is called a *black hole*. A black hole represents the highest density achievable in the cosmos. Once a critical mass is reached, then nothing will ever be able to stop a gravitational collapse, and the collapse will proceed to form a *singularity,* or a point of infinite density (in reality, when mathematical formulas representing real objects start spitting out infinities—it does not necessarily *literally* imply that something is large beyond measure, it means that the framework the physics and mathematics are based on have broken down and no longer give insight into the situation being examined). Black holes represent the highest densities in the cosmos outside of the big bang. While it is difficult to imagine these densities, you must keep in mind that matter consists mostly of empty space. Practically all of the volume that we associate with matter is merely the electromagnetic forces between atom's electrons maintaining a certain distance due to their mutual repulsion, and the strong nuclear force that keeps atomic nuclei at a stable size via gluons. The vast majority of matter is comprised of empty space that, under the right circumstances such as those presented in a black hole, can allow for incredible amounts of compression to occur.

Every object in fact has a certain radius that it can be compressed down to where its gravitational field will be strong enough to not permit light to escape, thereby becoming a black hole. Astronomers call this radius the *Schwarzschild radius* (named after its discoverer Karl Schwarzschild). For everyday objects, this radius is exceedingly small, but theoretically exists. This Schwarzschild radius beyond which not even light can escape, is also called the *event horizon*, since beyond this "horizon" we can no longer observe any "events" that occur since the photons of light (the information "source" for the event occurring) are bent back towards the singularity at the center of the black hole, and will never be seen. Mathematically, the Schwarzschild radius of a black hole is given by:

$$r_{schwarzschild} = \frac{2GM}{c^2}$$

Where G is the universal constant of gravitation, M is the mass of the black hole, and c is the velocity of light. The following table outlines a few sample Schwarzchild radii for various objects.

	mass (kg)	schwarzschild radius in (km)	in meters (m)	in centimeters (cm)
Moon	7.16E+22	0.00000011	0.000106333	0.010633347
Mercury	3.28E+23	0.00000049	0.000487362	0.048736175
Earth	5.97E+24	0.00000886	0.008861123	0.886112279
Uranus	8.96E+25	0.00013292	0.132916842	13.29168419
Saturn	5.67E+26	0.00084181	0.841806665	84.18066654
Jupiter	1.90E+27	0.00281784	2.817837048	
Sun	1.99E+30	2.95		

1.4 solar masses	2.79E+30	4.14
3.0 solar masses	5.97E+30	8.86
10 solar masses	1.99E+31	29.54
1 million solar masses	1.99E+36	2953707.60

It is difficult to think of light as bending in the presence of gravity as it contradicts our everyday intuitions and experiences. This effect has in fact been observed and is in tight accordance with relativistic predictions. While *any* object with mass and a gravitational field bends light, in order to observe this effect directly we need a massive object, such our sun. Our sun is massive enough at 1.99×10^{30} kilograms to bend light rays enough for us to observe here on earth. This bending of light can be seen when a total eclipse of the sun occurs, since the vast majority of the sun's light is being blocked by the moon's shadow. This allows astronomers to see the faint light of background stars near the sun that would normally be blocked out by its glare. By accurately measuring the position of these background stars that are near the sun during the total eclipse, and again when the sun is not nearby to bend their photons of light. Astronomers have observed and confirmed this bending of light to a high degree of accuracy.

This bending of light is often confusing for people. Many people know that photons of light have no mass. However, they do have an equivalent mass by means of Einstein's famous $E = mc^2$. So if gravity pulls on objects with mass, how it is that a massive object can bend a light ray that represents itself as a mass-less photon? This is possible because any object with mass or energy, *bends* its surrounding space-time just like a heavy ball on a thin rubber sheet. The heavier the ball, the more the sheet (space-time) gets bent. If you imagine a very small ball (a photon of light) rolling very rapidly past a point near the heavier ball on our rubber sheet, its path will be *deflected* by the bending of the sheet from the heavy ball. Despite having no mass, the photons of light are merely following the geometry introduced by mass, thereby diverting their trajectories.

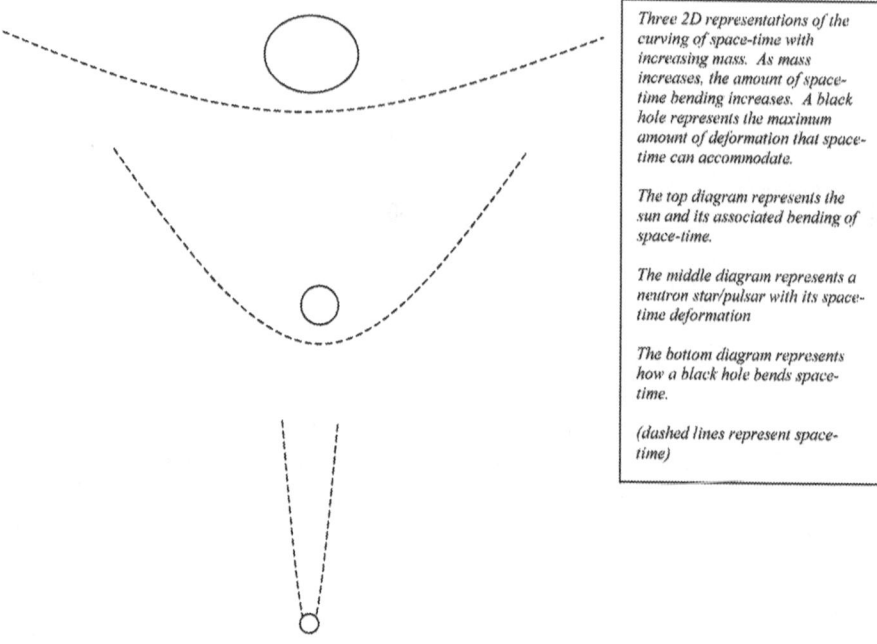

Three 2D representations of the curving of space-time with increasing mass. As mass increases, the amount of space-time bending increases. A black hole represents the maximum amount of deformation that space-time can accommodate.

The top diagram represents the sun and its associated bending of space-time.

The middle diagram represents a neutron star/pulsar with its space-time deformation

The bottom diagram represents how a black hole bends space-time.

(dashed lines represent space-time)

Three dimensional representation of space-time deformation resulting from a lower mass object, such as our sun.

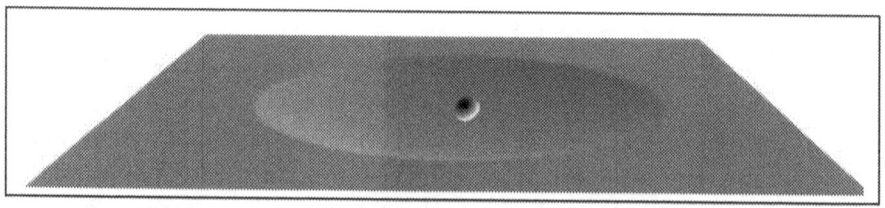

Three dimensional representation of space-time deformation resulting from a black hole.

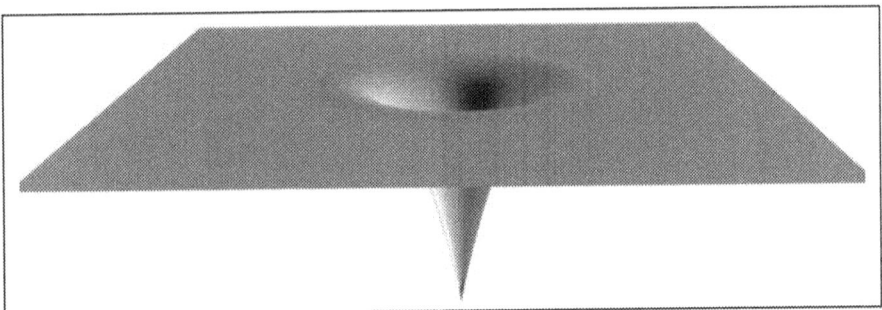

Another area where this bending of light rays by massive objects is seen is in an effect called *gravitational lensing*. This type of light bending has been seen in numerous areas of the sky where a fainter background galaxy's light (or group of galaxies) is amplified or *lensed* by a massive object(s) (just as a magnifying glasses convex spherical lens focuses light into a concentrated point) that are coincidentally in the path of their light. If the lensing object is lined up correctly, astronomers can observe what is called an *Einstein ring*, where the gravitational field of the intervening object "smears" the distant background objects almost entirely into a thin arc of gravitationally amplified light—approximating a ring shape. The detection of these objects is yet another validation of the concepts of general relativity.

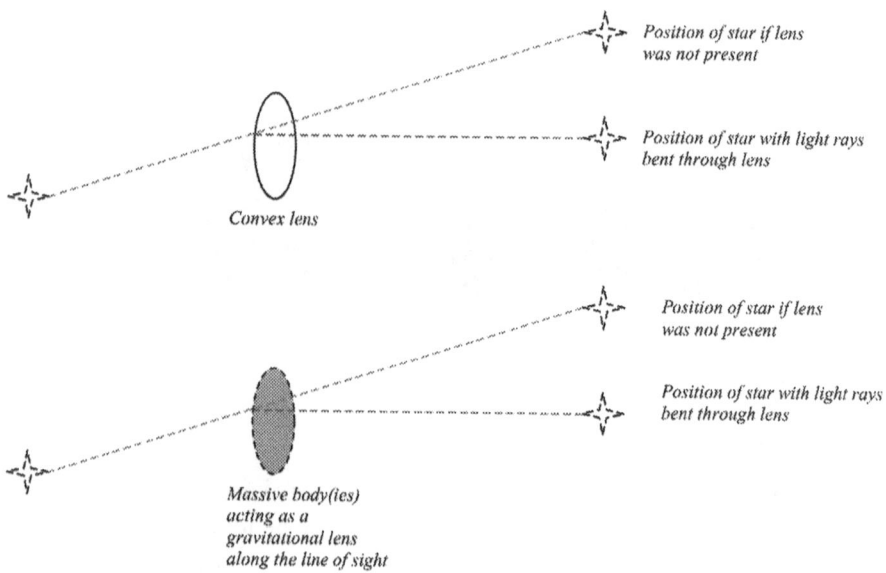

Position of star if lens was not present

Position of star with light rays bent through lens

Convex lens

Position of star if lens was not present

Position of star with light rays bent through lens

Massive body(ies) acting as a gravitational lens along the line of sight

Example of gravitational lensing. The top figure represents the refraction of a stars position as it passes through the denser material of a convex lens. The bottom figure represents an identical process, except that the convex lens has been replaced by a massive body (or bodies) whose gravity acts in a similar manner. Einstein's equations of relativity state that massive bodies will warp the surrounding space-time fabric nearby; and, since light must travel along the contours of these warps, it is warped as well and gets deflected as if the mass were acting as a lens. Numerous examples of this process of gravitational lensing have been found in astronomical photographs, providing yet more proof of Einstein's theory of relativity.

Despite the almost science-fiction type physics associated with black holes, their exploration mathematically goes all the way back to 1783, when astronomer John Mitchell first pondered what would happen to a star if its mass grew to a magnitude such that its escape velocity was equal to the speed of light.

5

Galaxies and Galaxy Types

Our Milky Way galaxy is one of *billions* of galaxies in the observable universe. It is not surprising that within all of the observed galaxies, there is a diverse range of properties associated with them. Despite this diversity, there is a relatively straightforward classification scheme used by astronomers to categorize the vast amounts of galaxies. This basic classification breaks down galaxies into four basic types based on appearance as first put forth by the astronomer Edwin Hubble (who also was the first to discover the expansion of the cosmos from making the first redshift measurements of galaxies outside the Milky Way), and as such is called the *Hubble classification scheme*. We will explore each of these four basic classifications of galaxies and the sub-categories associated with them in the sections to follow. It is important to keep an open mind with this (and any) classification scheme as it is by no means provides a hard and fast way to categorize the myriad of structures possible with galaxies—they are subjective.

While Edwin Hubble was the first to prove that the Universe is expanding, it was in fact the astronomer Vesto Silpher who found that galaxies were moving away from earth as seen by Doppler shifts in spectral lines of galaxies in 1912. The astronomer Willem de Sitter theorized that the universe could be expanding. Hubble met with William de Sitter in Holland and this then prompted Hubble to look for signs of cosmic expansion in galaxy spectral lines (<u>Parallel Words</u>, Pages 49 to 50).

Spiral Galaxies: Sa, Sb, Sc and Sd (Normal)

Spiral galaxies are, just as the name implies, spiral shaped. Our own Milky Way galaxy falls within this category. If you could peer down at the Milky Way from above, you would see a series of spiral "arms" that appear to be wrapped around a central "bulge". In the Hubble classification, all spiral galaxies are denoted by the letter "S". A further subset of letters exists within the "S" spiral galaxies as well. From observing thousands of spiral galaxies, astronomers found a relationship between the sizes of the central bulge of spiral galaxies, and how tightly "wound" their spiral arms appeared. This relationship provided the structure for the breakdown of spirals within the Hubble classification. Beyond the "S" for a spiral, there are Sa, Sb, Sc and Sd spirals. The a-b-c-d subset is simply reflective of the relationship between bulge size and spiral tightness mentioned earlier. Hence, an *Sa* spiral has a large central bulge and tightly wound arms that can almost be circular in shape. An *Sb* spiral galaxy would have a slightly smaller bulge than an Sa, and more "open" spiral arms. A *Sc* spiral would have a bulge that is smaller than a Sb's bulge, and also have a looser spiral structure which may have a more "clumpy" nature as well. The last type of spiral, the *Sd*, has a very small central bulge, and the loosest and most open and clumpy spiral structure of this class.

Barred Spirals: SBa, SBb, SBc (Normal)

Barred spirals are another subdivision of the Sa, Sb, Sc, Sd spiral classifications. Despite this subdivision, barred spirals are similar to spirals with the exception of their central bulges, which tend to be more smeared out into a shape that to some degree visually approximates a "bar". The rest of the distinguishing characteristics that categorize a regular spiral into Sa, Sb, Sc, or Sd grouping still applies to a barred spiral, we must merely add a "B" after the "S", which leads to: SBa, SBb, SBc, and SBd. Our Milky Way shows signs of having an elongated central bulge and may be a barred spiral as well (probably type SBc).

Elliptical Galaxies: E0 to E7, S0, and SB0 (Normal)

The characteristic difference between a spiral type galaxy and an elliptical type galaxy is in the presence or absence of spiral arms. *Elliptical galaxies have no distinguishable spiral arm structures.* As the name implies, the majority have a tendency to be spread out into an elliptical shape. For elliptical type galaxies we have *E0, E1, E2, E3, E4, E5, E6,* and *E7*. These 0→7 subdivisions are representative of the amount of eccentricity of the galaxies. An E0 elliptical galaxy is approximate to a circular shape, where an E7 is highly elongated.

As with any classification scheme, there are instances where an object's properties are straddled between two different classifications. In some instances, there are ellipticals that have hints of a galactic disk. Typically a disk like this is associated with a spiral type galaxy, and therefore falls into a sub-category of ellipticals, an *S0*. For ellipticals that have a slight galactic disk, and elongation to their central bulges, they fall under an *SB0* classification.

Below is an illustration of the Hubble classification scheme:

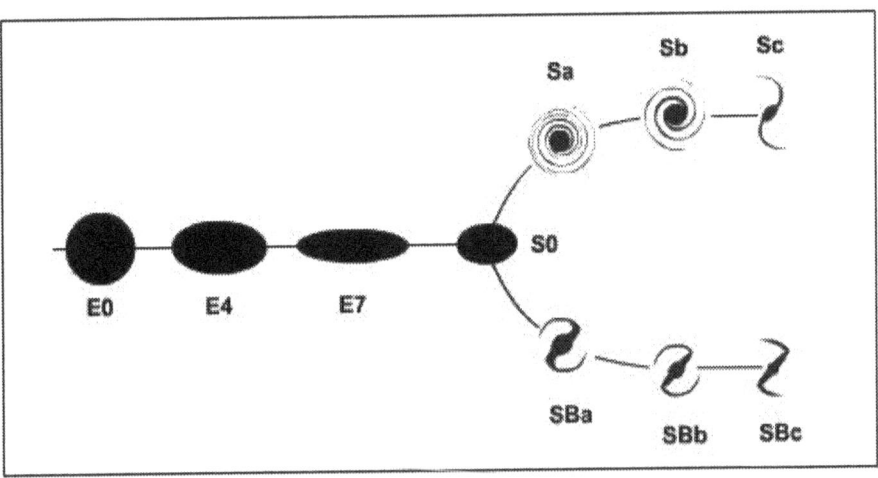

Irregular Galaxies: Irr I, Irr II (Normal)

Irregular galaxies simply represent the galaxies that do not fit into and spiral and elliptical classifications presented above. These are the remainders, and as one might guess, they typically have a lack of structure that would have placed them into a spiral or elliptical classifications. There are two types of irregular galaxies: *Irr I*, and *Irr II*. The first type of irregulars, Irr I often appear like spiral galaxies that were "interrupted" during their development. The Irr II galaxy types are not as common as their Irr I counterparts, and tend to have a filamentary structure or appear as if they are caught in the middle of an "explosion".

Active vs. Normal Galaxies

Our exploration of galaxy types now gets a bit more interesting. As we will see shortly, there are some galaxies that release copious amounts of energy and have much more "activity" than "normal" galaxies. These energetic and active galaxies are called just that—*active galaxies*. And, as for our everyday type galaxies that are not "active" (energetic) enough to fall under the active type, we have *normal gal-*

axies. All of the spiral, elliptical and irregular galaxy types that we covered to this point would fall into this *normal galaxy* family.

Another characteristic of active galaxies is their relative distribution. As you increase in distance, the number of normal galaxies tends to decrease while the number of active galaxies tends to *increase* (there are instances of normal galaxies at extreme distances, just as there are some instances of active galaxies at relatively close scales—but these are more the exceptions). Remember that the farther away a galaxy is, the farther back in time you are seeing as well. The light from a nearby galaxy 50 million light years away is already 50 millions years old by the time it reaches us (to reiterate—if somehow that galaxy somehow shut-off all its light output today—we would not know for 50 million years). Therefore if there is a tendency for more and more active galaxies to start popping up at larger and larger distances, this leads astronomers to believe that the early universe was a more energetic environment than we see today. Active galaxies are also character-ized by massive *jets* of high speed material being ejected from their nucleus, along with rapidly rotating clouds of matter spiraling around a compact energy source that can vary its energy output tremendously over a short period of time (most likely these phenomenon are caused by supermassive black holes in the galactic center on the order of millions of solar masses).

Seyfert Galaxies (Active)

The main difference between an active Seyfert galaxy and a regular spiral galaxy is found in their energy output. Based solely on appearance, Seyferts and spirals would appear very similar. However, once you start measuring the energy output, or luminosity, given the tremendous distances they are typically found at (most are several hundreds of millions of light years away), Seyferts are measured to expel *thousands* of times the energy output from a galaxy like our Milky Way. The majority of this energy output is concentrated in the central core regions of these galaxies (again—mostly likely fueled from matter falling into a supermas-sive black hole).

On top of having tremendous energy output from their core regions, Seyfert galaxies also show irregular variations in their energy output over time as well. A typical graph of the light intensity of a Seyfert observed over many years would look much like following a stock value on Wall Street. Since nothing is permitted to travel faster than the speed of light, whenever there is a focused variation in energy output from an object, you can multiply the time it takes for the variation to drop back down (let's say in years), by the speed of light to find quite confidently the size of the object producing the spikes (in light-years). In applying this line of thinking to Seyferts, we find that based on the intervals of their fluctuations, the object producing the variations in energy output from their cores are extremely compact, providing evidence that this is caused by supermassive black holes.

Most of this energy output is centered on wavelengths that fall in the radio or infrared part of the electromagnetic spectrum, and cannot be seen at visual electromagnetic wavelengths (there are fluctuations in other parts of a Seyferts spectrum, however the radio and infrared are the most salient). Radio and infrared wavelengths are not associated with the primarily visible wavelengths that stars or groups of stars emit their energy in. Therefore these radio and infrared emissions from Seyferts are most likely *not* associated with stars. Analyses of the spectral lines near the cores of Seyferts show evidence for gases that are rotating very rapidly around the center (nucleus) of the galaxy (on the order of several hundred to over one thousand kilometers per second).

Radio Galaxies (Active)

Another type of active galaxy in addition to Seyferts is *radio galaxies*. As the name implies, these galaxies have extremely high energy output in the radio region of the electromagnetic spectrum. They have comparable energy output in their central regions to that of Seyferts. In some instances, these radio galaxies have a small, compact central region, or "core" and an extended "halo" of weaker energy output surrounding it. These types of radio galaxies are referred to as *core-halo radio galaxies*. At radio wavelengths, these galaxies can emit a whopping 10^{37} Watts. With our sun's luminosity being 3.9×10^{26} watts, that means a core-halo radio galaxy expels about 256,410,256,410 *times* the energy output of our sun.

Radio lobe galaxies make up another subdivision of radio galaxies. Again, as the name implies, these galaxies have "lobes", or extended clouds of material (mostly gas) that are significantly beyond the central regions of the galaxy. These lobes emit a vast majority of their energy at radio wavelengths. The wattage output of these galaxies at radio wavelengths can approach 10^{39} watts. Radio lobe galaxies are among the largest observed in the cosmos, extending to several times the diameter of the Milky Way (some even approaching the size of our entire Local Group of galaxies).

Quasars—Quasi-Stellar Objects (QSO's)

At distances beyond Seyfert and Radio galaxies there is another class of active galaxies called Quasars, or quasi-stellar objects (referred to as QSO's as well). Quasars have a significant amount of overlap in the characteristics they share with our other active galaxies. Quasars often show signs of extended high-speed jets coming from high energy sources that are very compact. Despite the "stellar" appearance of many of these objects, they are in fact the brightest objects known so far in the cosmos. Their luminosities can approach a staggering 10^{42} Watts. This luminosity is the equivalent of *trillions* of suns or over 1000 times the luminosity of our entire Milky Way galaxy. The stellar appearance of these objects is merely a fact of their great distance, and their compact size. Quasars comprise the most distant objects observable in our cosmos, and provide a snapshot of the tremendous amounts of energy that were available in the early universe.

Quasar emissions have been detected at all wavelengths of the electromagnetic spectrum, from radio waves all the way up to gamma rays. As with our other active galaxies, Quasars also show rapid changes in their energy output over short spans of time. Even over a matter of *hours*, some Quasars can significantly increase or reduce their luminosities at certain wavelengths of the electromagnetic spectrum. Essentially, a Quasar is a Seyfert or Radio galaxy that is just a degree of magnitude more *active*. These objects, along with their other active galaxy cousins represent the more primordial view of the universe—the universe when it was much younger, hotter, and apparently more violent.

Black Holes Everywhere

So where are these active Seyfert and Radio galaxies from our last section getting all of this energy? They have such copious energy outputs compared to the other normal galaxies. From our discussion of Seyferts, we remember that they are found to have a very compact (small radius) source for their spikes of energy output, and, around these compact sources we have rapidly rotating high-energy gas. So, you have massive amounts of energy coming from extremely small volumes of space, with clouds of gas rapidly spiraling around it. For such magnitudes of energy to be released, we must look beyond our regular pictures of galaxies. Stars, even in massive quantities cannot produce enough energy output to fit the observations.

Based on this, practically all modern astronomers would agree that the culprit is a massive black hole at the center of these galaxies. To explain the raw energy numbers that we are able to observe from them, despite their tremendous distance from us (remember—active galaxies increase in frequency significantly with distance), black holes represent the best fit as the driving force behind these observations based on our currently available physics.

As with the stellar black holes that we covered earlier, the same processes are at work in active galaxies, except on an exponentially larger scale. The black holes we previously covered represented the final outcome of a single star, where, in active galaxy cores to account for the energy outputs that we are seeing, we must be dealing with black holes that are *millions* of solar masses or *billions* of solar masses. Such objects are hard to imagine—however, based on the evidence at hand, it is very likely to be accurate.

When large amounts of material are rapidly rotating and spiraling towards a center of gravity, we have an *accretion disk*. If, at the center of this accretion disk, we have a rotating supermassive black hole that contains the mass of billions of stars compressed into an area that is only a few light years across, where the local density bends space-time back onto itself so that light itself cannot escape its gravity, then you have an energy source that can produce the numbers we need (this process of accretion onto a rotating body is common on many scales of astronomy). As it turns out, if you have a massive enough object at the center of these accretion disks, they are actually more efficient at producing energy than nuclear fusion, having several times as much energy per unit mass as fusion.

In accretion at this scale, we need to think about entire stars and huge amounts of gas spiraling rapidly towards the black hole with the equivalent mass of an entire galaxy. As the mass of stars and gas in the accretion disk spiral towards the black hole, the temperatures begin to rise rapidly due to friction. With this increase in temperature comes a corresponding increase in emitted energy. As objects get close to the supermassive black hole, they are ripped apart all the way to the atomic level and heated to immense temperatures. As these extreme energies, strong magnetic fields develop and charged particles stream out perpendicular to the plane of the accretion disk along these lines, producing the massive jets that are sometimes observed. The faster these charged particles are moving, the more intense the reinforcing of the magnetic field and hence the more energy that is released. Charged particles get tangled up in these strong magnetic field lines, spiraling around them on their ejection paths from the accretion disk. This type of radiation is called *synchrotron radiation*. Since there are variations of density within the accretion disk surrounding the supermassive black hole, there are "intervals" or "bursts" of rapid accretion over short periods of time as more dense clumps of the accretion disk come into the proximity of the black hole. This explains the large variations in energy output that we observe.

So now that we have a good idea what is happening in the centers of active galaxies to produce their unusual properties, what about the normal galaxies that are closer to home? How "normal" are they? As it turns out, their "normal" classification hides a secret. As astronomers began to peer into the cores of normal galaxies in our cosmic neighborhood, they noticed the same types of basic properties of active galaxies' cores, but on smaller scales. Practically all normal galaxies cores show evidence of rapidly rotating masses of stars and gas clouds around compact sources of mass. In a nutshell, it is appearing increasingly likely that *all* galaxies house supermassive black holes in their centers—including our own Milky Way. It seems that central black holes may be part of the normal formation and evolution of galaxies.

If we step back a minute this actually makes sense. As I have mentioned previously, the farther away objects are from us, the farther back in *time* we are seeing them due to the accelerating expansion of the cosmos and the unsurpassable speed of light discovered by Einstein. The active galaxies (Seyfert, Radio, and Quasar) we covered earlier are found in much greater densities as you increase distance. This increase in distance corresponds to an earlier time in the cosmos. Many astronomers have put forth that active galaxies are merely the earlier evolutionary counterparts of normal galaxies that we find at lesser distances. If early, active galaxies all gathered up and accreted enough matter to curve space-time into supermassive black holes in their cores, then naturally they would be found in our close normal galaxies as well. The reason for this is that black holes do not simply vanish over time. They have small amounts of mass that they emit away over time (called black hole *evaporation*—see section below), however the time scales calculated for this to occur are far beyond the current age of the cosmos, so the black holes would remain in the centers of active galaxies that calmed down over time to become normal galaxies. Much of the raw material that was being accreted onto the supermassive black holes in the centers of active galaxies has already been cannibalized and used up by the time a galaxy has evolved into a normal type galaxy. The black hole remains in their centers; however with the absence of material accreting on to them, we do not directly see their presence as we do in the younger active galaxies where stars and gas clouds would be accreting onto their surfaces much more frequently with the violent energy emissions observed. With these supermassive black holes, we are dealing with objects as massive as *billion* of suns, squeezed into the smallest area possible that is allowed by our current understanding of physics.

Evaporation of Black Holes

Indeed, black holes are very odd and mind-boggling objects. In addition to their ability to bend space-time to such a degree through their gravitational fields so as to prevent photons of light from escaping, they also *emit* radiation as well. How can this be? If a black hole's gravity is so powerful, how can it emit radiation, when that radiation is made up of photons? Stephen Hawking was discoverer of this radiation that black holes emit, and it is referred to as *Hawking radiation.* To explain the details of the evaporation of black holes, I will quote an excerpt from Stephan Hawking's book <u>The Theory of Everything: The Origin and Fate of the Universe (pages 82-84)</u>: "What we think of as empty space cannot be completely empty because that would mean that all the fields, such as the gravitational field and the electromagnetic field, would have to be exactly zero. However, the value of a field and its rate of change with time are like the position and velocity of a particle. The uncertainty principle implies that the more accurately one knows one of these quantities, the less accurately one can know the other." We will cover the uncertainty principle and the position and velocity oddities that occur in quantum physics later in this book.

The quote continues: "So in empty space the field cannot be fixed at exactly zero, because then it would have both a precise value, zero, and a precise rate of change, also zero. Instead, there must be a certain minimum amount of uncertainty, or quantum fluctuations, in the value of a field. One can think of these fluctuations as pairs of particles of light (photons) or gravity (gravitons) that appear together at some time, move apart, and then come together again and annihilate each other. These particles are called virtual particles. Unlike real particles, they cannot be observed directly with a particle detector. However, their indirect effects, such as small changes in the energy of electron orbits and atoms, can be measured and agree with the theoretical predictions to a remarkable degree of accuracy." This remarkable degree of accuracy is the equivalent of estimating the distance between two cities thousands of miles apart, to within the thickness of a typical sheet of paper! Quantum theory is the most accurate theory ever conceived. Back to our quote: "By conservation of energy, one of the partners in a virtual particle pair will have positive energy and the other partner will have negative energy. The one [particle] with negative energy is condemned to be a short-lived virtual particle. This is because real particles always have positive energy in normal situations. It must therefore seek out its partner and annihilate it. However, the gravitational field inside a black hole is so strong that even a real particle can have negative energy there.

It is therefore possible, if a black hole is present, for the virtual particle with negative energy to fall into the black hole and become a real particle. In this case it no longer has to annihilate its partner; its forsaken partner may fall into the black hole as well. But because it has positive energy, it is also possible for it to escape to infinity as a real particle. To an observer at a distance, it will appear to have been emitted from the black hole. The smaller the black hole, the less far the particle with negative energy will have to go before it becomes a real particle. Thus, the rate of emission will be greater, and the apparent temperature of the black hole will be higher.

The positive energy of the outgoing radiation would be balanced by a flow of negative energy particles into the black hole. By Einstein's famous equation $E=mc^2$, energy is equivalent to mass. A flow of negative energy into the black hole therefore reduces it mass. As the black hole loses mass, the area of its event horizon gets smaller, but this decrease in the entropy of the black hole is more than compensated for by the entropy of the emitted radiation, so the second law [of thermodynamics] is never violated."

For reference, the second law of thermodynamics states that an isolated system will always move from lower entropy to higher entropy, where the term *entropy* is the measure of disorder in a system (If you put your dirty car in your garage, it does not become more clean the longer you leave in your garage—it gathers more dust and becomes more dirty). A black hole represents the highest entropy possible. The main reason for wanting to explore this evaporation of black holes more thoroughly is to point out that they are active objects despite their names. This evaporation process is extremely slow—as Stephan Hawking states "even a black hole with a mass a few times that of the sun….would take about 10^{66} years to evaporate completely."

Once a black hole has completely evaporated, it is thought that it will release enormous amounts of energy in a huge explosion. The reason for this theory is the law of the conversation of mass and energy. If a black hole has been gobbling up mass for its whole life until it evaporated away, it must release a tremendous amount of energy to balance out what it consumed during its existence. Only when this excess energy has been released will the conversion of energy be complete. After this massive energy is released, only then will the black hole be "allowed" to snap out of existence.

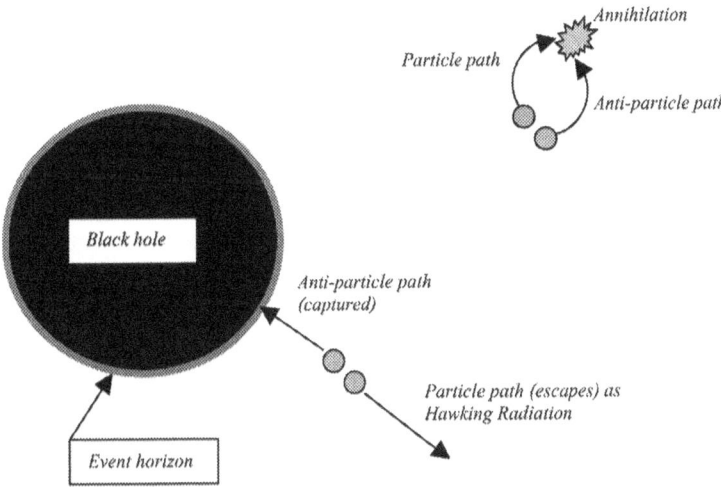

Black hole evaporation based on the production of virtual particle pairs from the vacuum energy of space. The quantum uncertainty principle dictates that there must be a fundamental uncertainty in the energy level of the vacuum of space. Due to this uncertainty, there can be enough energy based on the mass-energy relation to create a particle and anti-particle pair. These so-called virtual particles usually recombine and annihilate each other back into pure energy. Near the event horizon of a black hole, it is possible for some virtual anti-particles to get captured by the black hole and annihilate with regular matter within the black hole. This annihilation into pure energy within the black hole causes a reduction in its mass—again in accordance to the mass-energy relationship. The other particle in the pair will then escapes and becomes part of the Hawking radiation for that black hole. In this manner, a black hole slowly will "evaporate" over large enough time-scales.

Primordial Black Holes (mini black holes) in the Early Universe

As we saw earlier, "even a black hole with a mass a few times that of the sun…would take about 10^{66} years to evaporate completely." If it takes such a long time for a black hole to evaporate through the emission of Hawking Radiation, then why do we care about this process? In part, it is significant because in the early universe it has been proposed that many small-mass black holes may have been forming in considerable numbers from the increased density of matter shortly after the big bang if there were enough irregularities in the matter distribution at that early epoch. These *primordial black holes* or *mini black holes* would be small enough in mass, and have a large enough Hawking Radiation from evaporation to be observed. As John D Barrow explains in his book, <u>The Book of Nothing (pages 227-228)</u> "In order for Hawking's radiation process to be visible, we would have to encounter black holes which are only about the mass of a large mountain or asteroid. Their [event] horizon size is equal to that of a single proton! These 'mini' black holes cannot form today when stars die. But they can be formed in the dense environment of the Big Bang if it is irregular enough. If there were, then these mountain-mass black holes would be in their final stages of evaporation today. The climax of this process will be a dramatic explosion that would show up as a burst of high-energy gamma rays accompanied by radio waves arising from the fast-moving electrons emerging from the explosion at speeds close to that of light. They would radiate 10 *gigawatts* of gamma-ray power for a period of *forty billion years* and could be seen many light years away. Radio telescopes could see the radio waves from one of these atomic-sized explosions occurring two million light years away in the Andromeda galaxy." Here in lies the significance; if we can detect these tiny primordial black holes as they explode out of existence, then we can learn a tremendous amount of information about the early history and evolution of the universe. Efforts have been made to detect these predicted explosions, however as of this printing, none have been found. This lack of observational evidence does not prove or disprove their existence, it could merely mean that they were less frequently formed early in the cosmos (which means the early universe has less irregularities) or that many of these primordial black holes were smaller in mass and have already evaporated, exploded, and cease to exist.

When higher energy particle accelerators come online, it is hoped that they might have enough energy to create tiny black holes, which would immediately evaporate away and explode (not cataclysmically), thereby producing a shower of other particles that would be detectable. According to quantum physics based in our three spatial dimensions, the next line of accelerators will still be shy of the energies needed to produce these quantum-sized black holes. However, if the energy calculations are done in higher dimensional spaces as those we shall explore later in superstring theories, and m-theory, then the required energy drops to a level that might allow these new devices to create them, bringing forth evidence for the validity of higher dimensional physics (there is a good article in the May 2005, Scientific American that outlines these concepts in greater detail).

So as it turns out, once of the most profound and powerful objects known in the universe, a black hole, is turning out to be a very common object in our cosmos. Perhaps in the future, new physics and mathematical theories will be discovered that can shed "light" on black holes. Until then however, we will never be certain of what happens on the other side of the event horizon. Despite this lack of information on what really happens to matter falling into the depths of a black hole and its accompanying singularity, there are theories on what may take place.

Dark Matter Everywhere

Once the technology for analyzing spectral lines and their associated shifts was linked to radial velocities, it started to become apparent that there was an abundance of unseen or *dark* matter that permeated the cosmos. In the 1930's, astronomer Fritz Zwicky was studying the Coma cluster of galaxies. When he measured the dynamics of how the individual galaxies within the cluster were moving, he soon realized that their measured velocities were much too great for the cluster to have not dispersed. In order for the measurements to be true (without altering the laws of physics) he realized that there would have to been a large unseen type of mass that was holding the cluster together.

In 1962, astronomer Vera Rubin noticed the same types of problems as Zwicky when analyzing the rotational velocities of the Milky Way. There was simply not enough gravity with the associated visible matter in the Milky Way to explain how it could be rotating so rapidly without being ripped apart. Since galaxies are obviously longer term features of the cosmos that last for billions of years there must indeed be some hidden dark matter. When Rubin crunched the numbers for how much of the Milky Way's matter is dark matter, she came up with a staggering 90 percent. After finding such startling results in the Milky Way, Rubin soon pursued the same types of measurements on other spiral galaxies, and came up with similar results. Galaxies were spinning too fast to be held together by their detectable mass.

While the exact composition of this dark matter is currently unknown, theories abound. Some theoretical physicists envision dark matter has being some form of weakly interacting massive particles, or *WIMPS*. It is a source of both frustration and amazement that such a large percentage of the mass of our cosmos is invisible and undetectable.

6

Einstein's Special Relativity

Einstein was an immense intellect, and his work on the theories of special and general relativity represent perhaps the most demanding intellectual discovery that has every been revealed by a single mind. The vast majority of the key principles behind these theories were derived by Einstein himself, with little collaboration among other physicists and mathematicians. In our modern era, relativity theory has become table talk—popular knowledge of sorts. However, to get a true understanding of Einstein's theories and equations requires much diligence. By no means is this possible in this, or any single text. However, I do want to pick up a large brush from our intellectual toolbox and paint a broad picture of the essentials of special and general relativity. I hope that this will ignite an interest in the topic. Any reader will greatly profit from further reading on the topic of relativity. What is amazing beyond all of this is the fact that such a ground breaking discovery such as special relativity was unveiled in 1905. To this day, the implications of Einstein's special and general relativity theories are still being revealed. The year of 1905 is known as Einstein's "miracle year" since in addition to publishing his results on special relativity, he also published his results for Brownian motion, the photoelectric effect, and the equivalence of mass and energy in the famous $E=mc^2$ relationship. Surprisingly, Einstein ended up winning the Nobel Prize for his work on the photoelectric effect rather than special or general relativity.

Newton's second law of motion is familiar to all of us. It states that the Force an object exerts is equal to its mass multiplied by its acceleration. In essence, relativity is merely a modification of this law that takes into account the increase in mass that occurs when an object is moving. How can an object increase in mass when it is moving? Such a concept is not as far-fetched as it sounds. In Richard Feynman's classic book <u>Six Not-So-Easy Pieces</u> (page 49-50) he gives a good "in a nutshell" definition of the concept of relativity: "Newton's Second Law which we have expressed by the equation

$$F = \frac{d(mv)}{dt} \text{ , [or better known as } F = ma \text{],}$$

was stated with the tacit assumption that m is a constant, but we now know that this is not true, and that the mass of a body increases with velocity. In Einstein's corrected formula m has the value:

$$m = \frac{m_o}{\sqrt{1 - v^2/c^2}} \text{ ,}$$

where the "rest mass" m_o represents the mass of a body that is not moving and c is the speed of light, which is about 3×10^5 km/sec or about 186,000 mi/sec [a more exact value for c is 299,792,458 km/sec]...For those who want to learn just enough about it so they can solve problems, that is all there is to the theory of relativity—it just changes Newton's laws by introducing a correction factor to the mass." Based on this information, Newton's Second Law becomes:

$$F = \left(\frac{m_o}{\sqrt{1 - v^2/c^2}} \right) \cdot a$$

While the above statement is perfectly valid, we must keep in mind that the velocities that we encounter on our everyday dealings are nowhere fast enough to produce a noticeable difference in our mass. I mentioned earlier in the book that technically you are more massive when you are in motion say, on an airplane. While that is a truth, the amount of mass increase is less than one part in a *billion* for the low velocities of an airplane. Even if you took a mass of 1 kg at rest on the surface of the earth, and enclosed it in a powerful rocket that reached a top speed of 100 km/sec, from the frame of reference of the ground, you would only notice a tiny increase of 0.00000000000006 kg.

Special Relativity Restrictions

It should be noted that all of our discussions of Special relativity need to be placed into a certain type of framework, mainly, that there are only certain types of situations where we can use special relativity. First, special relativity assumes that all the motions are uniform straight-line motions. That is, that there are not changes in direction or velocity that would be interpreted mathematically as acceleration. *No accelerations are allowed in special relativity.* Second, *special relativity assumes that the object is not rotating as it is moving.* Third, *special relativity assumes that the conditions of uniform non-rotational velocities are occurring in the absence of a gravitational field.* It should also be noted for clarification that while the speed of light is constant in a vacuum, in reality materials such as water, glass, and air affect the rate of lights speed—hence why we see such everyday effects such as iridescence, refractions, and scattering.

Frames of Reference

At this juncture, it is good to explore how different frames of reference can change how a particular event unfolds. By frames of reference in relativity speak, we mean different rates of motion as seen by different observers. In our above airplane example, the ground is unmoving *relative* to the airplane so the ground represents a *separate* frame of reference. Likewise, from inside the airplane looking out a window, the plane seems unmoving (if it is not accelerating—then you would feel it), and the earth seems to be in motion, hence the plane is also a *separate* frame of reference as well *relative* to the earth. We can easily expand these frames of reference to different scales. For instance, from the earth's frame of reference, the sun appears to be moving. From the sun's frame of reference the earth is in motion around it while it is at rest. From the sun's frame of reference, the center of our Milky Way galaxy appears unchanging and static. From the Milky Way center, the reference frame shows our sun in orbit motion around it. If we placed a reference frame on a nearby galaxy, it would show that our entire galaxy is orbiting the center of mass of the Local Group...etc. We can continue this game forever...or can we? If we imagine larger and larger scales, picking frames of reference as we go, we eventually reach the level where the reference frame is the entire visible universe. This is an interesting matter to investigate, for if everything inside the visible universe is in apparent motion from various reference frames, then if the universe *itself* is in motion—from what frame of reference it is in motion? We will address this question shortly; first however, we will profit from some background.

The Principle of Relativity

A basic premise behind the theory of relativity is that all the laws of nature that apply for one frame of reference, can be applied to another frame of reference. More specifically for our purposes here, this means that frames of reference with different apparent motions with respect to each other still follow the same laws of nature—or that nature's laws are equivalent in different frames of reference. According to the best of our mathematical knowledge and experiments, this is indeed true. In Einstein's own words from <u>Relativity—The Special and General Theory</u>, page 61, he describes the principle of relativity as it applied to special relativity as: "All bodies of reference K, K'...are equivalent for the description of natural phenomena (formulation of the general laws of nature), whatever may be their state of motion."

The Michelson-Morley Experiment

In the past, people thought of objects as moving through *aether* that surrounded all matter as it traveled in space. The concept of the aether was introduced due to the fact that photons can propagate in the vacuum of space. The aether was created to explain what was doing the "waving". Just as air molecules moving produces sound waves and water molecules moving produces water waves, the aether was the medium through which light traveled in the vacuum of space. This aether was thought of as a fine mist that had practically no viscosity and would be difficult to detect experimentally. In 1887 Michelson and Morley performed an experiment which was sensitive enough to measure the effects of this proposed aether. This experiment utilized a light source that sent a beam of light to a *partially* reflective mirror that was tilted +45 degrees. This splits the beam into two parts, one which travels upwards perpendicularly to the original line of travel and another that continues undeflected through the mirror. At *equal distances* from this partially reflective mirror are two other fully reflective mirrors, one which lies along the straight path of the light beam to bounce back the light that did not get reflected by the tilted partially-reflective mirror, and another that lies perpendicular to bounce back the parts of the beam that did reflect back from the tilted partially-reflective mirror. These two separate beams then bounce back to the tilted mirror and get *recombined*, with some of them then getting reflected perpendicularly *downward*, towards a detector. Here is the key point: if this apparatus is firmly mounted to the earth, which is in motion relative to the aether, creating a "flow" of aether in the direction of the earth's motion, if one of the light beams paths encounters some resistance or "drag" from this aether, then it will have traveled for a slightly different time since the apparatus is on the moving earth. This difference in time could be detected by the interference pattern of the combined light beams, since different times mean that the beams would be out of phase with respect to each other. This would show up in the interference patterns of the recombined beams. By rotating the apparatus at different angles and observing the interference patterns, the aether could be detected. This experimental apparatus was originally suggested by James Clerk Maxwell, the father of electromagnetism, as a way to decisively test for the existence of the presumed aether (which he believed to exist, as did most physicists of the time).

So what were the results of this clever experiment? Simply put—nothing. Regardless of what angles the experiment was oriented, there were no observed interference patterns, and therefore no differences in the times, and hence *there was no experimental proof for the aether*. While other physicists attempted to explain away the lack of results for the existence of the aether, Einstein instead proceeded on the simple premise that the aether could not be experimentally validated because it does not exist.

Relativity of Simultaneity

Now that we have covered the concept of frames of reference, and their implications on events, let us now explore the concept of the *relativity of simultaneity*. We will employ the use of example to illuminate some key points. Let us place ourselves in a car that is traveling at a constant velocity v along a straight section of highway. This is no ordinary car in the sense that it only has one window—the windshield—so we are limited to observe only what is in front of the car. Our friend, Sam, is sitting in a coffee shop that is a few feet off the highway that we are traveling on. As Sam is sipping her coffee, she notices a remarkable event: two bolts of lightning striking two distant, *widely separated* trees along the edge of the *same* side of the highway—*simultaneously*. In our car traveling down the highway at uniform velocity v, we observe the same remarkable event. What are the differences in this event from our frame of reference in the car, and Sam's frame of reference, who sits motionless in the coffee shop? At first thought, this may seem like a silly example, however with further thinking, a discrepancy becomes visible.

From Sam's motionless frame of reference, in order for the events to happen simultaneously, that means that the light from both of the lightning bolts reached her eyes at precisely the same time. That light is traveling at a constant velocity as well—c (about 3×10^5 km/sec). In order for simultaneity to occur, the light from the tree farthest from Sam must have started a tiny bit *earlier* than the light from the nearest tree being struck in order for them to reach her simultaneously, because it had to travel a slightly longer distance. From our moving car, we do *not* see a simultaneous event. Our velocity v, in the car that is *towards* the trees being struck changes the distance that the light from the two bolts need to traverse to reach us—breaking down the simultaneity of the event. This is the principle behind the *relativity of simultaneity*. In essence, it states that different frames of reference have different *times*. *Time is nothing other than the measurement of relative motions, and all of the measurements of the motions associated with the passage of time are attached to an individual frame of reference.* You cannot isolate yourself from a frame of reference.

Another example involving trains and lightning, in Einstein's own elegant language, can be found in <u>Relativity—The Special and the General Theory</u> on pages 25-27: "We suppose a very long train traveling along the rails with the constant velocity v....People traveling in this train will with advantage use the train as a rigid reference-body (co-ordinate system); they regard all events in reference to the train. Then every event which takes place along the line also takes place at a particular point of the train. Also the definition of simultaneity can be given relative to the train in exactly the same way as with respect to the embankment. As a natural consequence, however, the following question arises:

Are two events (e.g. two strokes of lightning A and B) which are simultaneous *with reference to the railway embankment* also simultaneous *relatively to the train?* We shall show directly that the answer must be in the negative.

When we say that lightning strokes A and B are simultaneous with respect to the embankment, we mean: the rays of light emitted at the places A and B, where the lightning occurs, meet each other at the mid-point M of the length A to B of the embankment. But the events A and B also correspond to positions A and B on the train. Let M' be the mid-point of the distance A to B on the traveling train. Just when the flashes (as judged from the embankment) of lightning occur, this point M' naturally coincides with the point M, but it moves...with the velocity v of the train. If an observer sitting in the position M' in the train did not posses this velocity, then he would remain permanently at M, and the light rays emitted by the flashes of lightning A and B would reach him simultaneously, i.e. they would meet just where he is situated. Now in reality (considered with reference to the railway embankment) he is hastening towards the beam of light coming from B, whilst he is riding on ahead of the beam of light coming from A. Hence the observer will see the beam of light emitted from B earlier than he will see that emitted from A. Observers who take the railway train as the reference-body must therefore come to the conclusion that the lightning flash *B* took place earlier than the lightning flash *A*. We must arrive at the important result:

Events which are simultaneous with reference to the embankment are not simultaneous with respect to the train, and *vice versa* (relativity of simultaneity). Every reference-body (co-ordinate system) has its own particular time; unless we are told the reference-body to which the statement of time refers, there is no meaning in a statement of the time of an event."

Obviously this example is to be taken with a grain of salt, and is a thought experiment for demonstration purposes only. In reality there are variations on the speed of light in air as opposed to in a perfect vacuum, and the human eye or mind is by no means fast enough to capture such discrepancies on the order of billionths of a second.

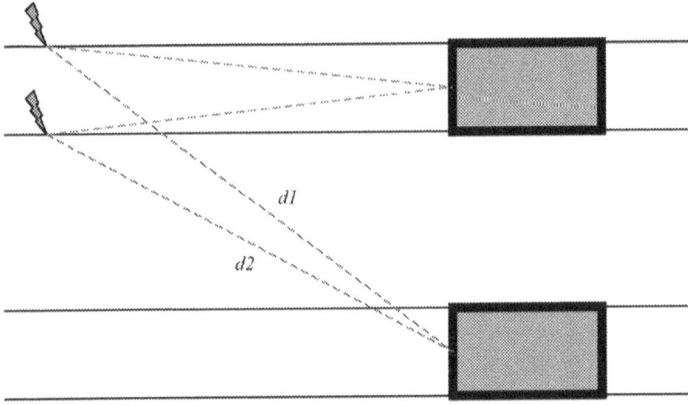

Illustration of the relativity of simultaneity. The box car in the top figure observes two bolts of lightning that strike the track it is on at exactly the same time as observed from a detector housed at the front of the box car. The photons from the two lightning bolts reach the detector in the top box car at exactly the same time. From the bottom box car's frame of reference, the lightning strikes were not simultaneous as the photons represented by d1 had to travel a slightly longer distance to reach the detector in the bottom car, as compared to the photons of d2. This is the concept behind the relativity of simultaneity. Simultaneous events are dependent on the observer's frame of reference.

We can apply another example to this as well. Let us take two clocks and place one at the front of the car, and one at the rear. We then measure the exact mid-point of the car and place two photon guns that will fire towards the clocks at the front and rear of the car. We have set our clocks to only start ticking when they receive a photon from the guns at the midpoint between them. What will the result of this be to us, the passengers in the car? For us, we will see both photons go out and hit the clocks at the front and rear of the car at the exact same time, making the clocks tick in perfect synchronization, ticking simultaneously. What about Sam? Will she see our clocks synchronized, ticking simultaneously as well? She will not. From her frame of reference outside of the car, our clocks are both in motion, so therefore the photon that was fired out of the gun towards the clock in the front of the car has to travel a greater distance than the photon fired towards the clock in the rear of the car. The reasoning is that the clock in the front of the car is speeding *away* from the photon at the velocity of the car, so the photon has to cover the half the distance between them *plus* the extra distance that the clock moved in the time it took the photon to get there. Conversely, the clock in the rear of the car is speeding *towards* the oncoming photon at the velocity of the car, so the distance that the photon needs to travel to activate that clock is half the distance *minus* the distance the rear clock moved forward while the photon was traveling there. Both of the above are examples of *failure of simultaneity at a distance.*

Dynamic Time—Dilation

The discovery that simultaneity is dependent on the frame of reference, more specifically that each frame of reference has its own time, leads us to another interesting part of relativity: *dynamic time, or time dilation.* No longer is time allowed to be considered universal as in Newton's eyes. Time is motion, and that motion (and hence time) is subject to a certain degree of plasticity when dealing with different frames of reference that are in different apparent motions. As physicists would say, time *transforms* in different frames of reference to accommodate the speeds at which frames of reference are moving relative to each other. This ensures that nothing can travel faster than light, and that you will never catch up to a racing light beam. To place this concept in a more formal form, if an observer is stationary relative to what they are *observing*, the time t that goes by for them is t. To the motionless observer, one second is one second—one hour is one hour on their clock. Let us take our example from before where Sam was motionless in the coffee shop while we were in our car speeding down the highway. From Sam's frame of reference, her clock is running at normal speed. *To us in the car, our watches are also running at normal speed since we are traveling at the same velocity as our frame of reference—the car.* If we were to beam the information from our moving clock out of the car, back to Sam's spot, she would notice that our clock is running a bit slow, since as we speed by, any information from our clock can only be communicated at the speed of light, so inevitably there is slight delay, or slowing of time on Sam's end. In a more formal description the formula for a unit of time is given by:

$$t = \frac{1}{\sqrt{1 - v^2/c^2}}$$

Where t is the time as measured by the observer at rest, v is the velocity of the moving object, and c is the speed of light (exactly 299,792.458 km/sec). It should be noted that this equation is set up for the special circumstance that the motion is uniform, restricted to one direction, and is non-rotating. In reality, an object can move in the x,y, and z directions in different combinations and this adds considerable complexity to the mathematics involved, so this is an idealized case to grasp the concepts.

Returning to our example, let us consider that we have a simple clock that consists of a photon gun that shoots out a photon of light from a point in the car, bounces off a mirror a set distance away, which reflect the photon back down to a detector that is nearby. Each time a photon hits the detector after making its journey, it makes what we consider a "tick" of our clock. In our frame of reference we don't notice anything unusual as we are traveling in the car at the same velocity as the car and its accompanying clock. In the coffee shop, Sam has an identical type of clock; however she notices that our clock is "ticking" more *slowly* than hers. Why is this? The solution is in the fact that from Sam's frame of reference separate from the car, each time a photon leaves the photon gun and bounces off the mirror on its way to the detector it has to travel a *longer* distance to reach the detector since the whole clock has *moved* under the photon with the velocity of the car while the photon was traveling. This longer distance corresponds to a longer time for the photon to get to detector to make a "tick" in our clock as seen from Sam's perspective. Again, to us in the car, we notice nothing unusual as we are traveling at the same velocity as our clock; however, to us, Sam's clock would appear to be slower, since relative to our car, Sam's clock is in motion and causes the same effect to be seen by us. The above principle is the same as what was explained in the Michelson-Morley experiment as well. What this implies is extraordinary: each frame of reference has its own unique time attached to it that is built into the laws of nature. Newton's notion of a global concept of absolute time, regardless of frame of reference was wrong. The table below outlines the time as represented by a motionless frame of reference, and a reference frame that is in motion (note the dramatic increase in time as the velocity approaches the speed of light).

Time elapsed for stationary observer (seconds)	Velocity v (km/sec)	Time of moving clock as viewed from a rest frame of reference (seconds)
1.00	0.1000000	1.00000000000006
1.00	100.0000000	1.00000005563251
1.00	1000.0000000	1.00000556329671
1.00	10000.0000000	1.00055678970520
1.00	100000.0000000	1.06075200044420
1.00	200000.0000000	1.34238470084142
1.00	250000.0000000	1.81192213751449
1.00	260000.0000000	2.00866169714504
1.00	270000.0000000	2.30096062455641
1.00	280000.0000000	2.79855957223183
1.00	290000.0000000	3.94480224624939
1.00	299000.0000000	13.76240756632980
1.00	299700.0000000	40.26767860535430
1.00	299791.0000000	320.63957165860400
1.00	299792.0000000	572.08792402792100
1.00	299792.4500000	4328.62899110104000
1.00	299792.4570000	12243.21168959590000
1.00	299792.4579000	38716.43932490720000
1.00	299792.4579900	122432.38584603300000
1.00	299792.4579990	387166.16512020500000
1.00	299792.4579999	1224418.61354763000000
1.00	*299792.458 = c*	*infinity*

Remember, time is nothing more than motions over distances. Units of time are not something that was created in the creation of the cosmos; they are human conceptions to aid in our understanding when analyzing physical systems. A better and more accurate way to envisage distances is in terms of time. Just as later we will discover that there is equivalence between energy and mass, you can also think of time and distances as interchangeable attributes. For example, you can think of 3×10^8 meters as a distance, or as approximately one second traveling at the speed of light. This is a handy concept, especially for astronomers, for it makes clear that everything we see happened at some point in the past since the information from any given object can only be sent towards us at light speed. In astronomy, as we discussed earlier in the book, this is why when astronomers gaze up at stars that are hundreds of light-years away, or galaxies that are millions or billions of *light-years* away, they are seeing them as they were hundreds, millions, or billions of years in the past since light emitted by them has a constant velocity (hence using the term light-years). There is no such notion as "now" in the literal sense. We are all living in a time machine of sorts, where the distance back in time is tied to the distance of the objects we are looking at in our world.

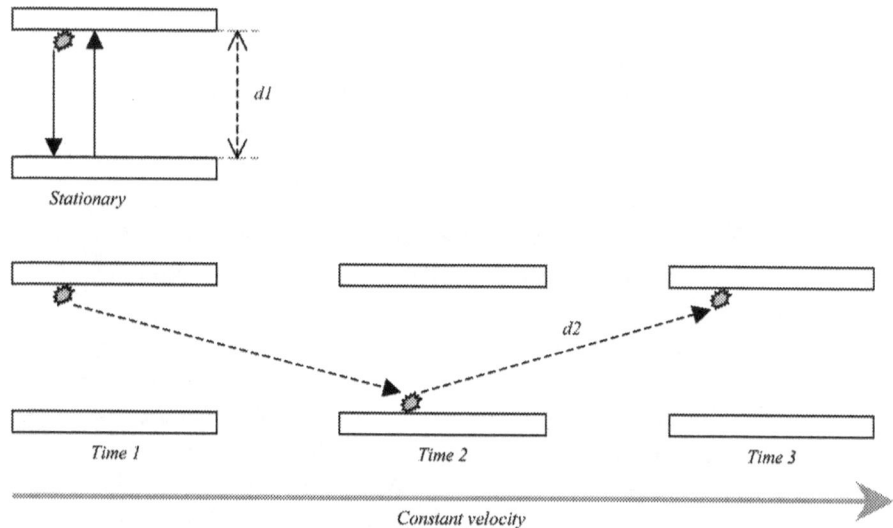

A simple representation of time dilation as a result of speed. The top figure represents a photon clock where each tick of the clock is represented by the photon traveling down from the top plate, bouncing off the bottom plate, and returning back to the top plate. If this clock is static, then the photon needs to travel twice the distance d1 to get back to the top plate. In the bottom figure, this same photon clock is moving at a constant velocity towards the right. When the photon leaves the top plate, the apparatus will move to the right by the time the photon hits the bottom plate. On the way back up to the top plate, the clock has moved once again from the velocity. This movement increases the distance that the photon needs to travel to record a tick, thereby making the clock appear to run slower to an observer outside the reference frame. The faster the clock is moving, the more movement the clock will make between ticks and the slower each tick will be, since the distance d2 that the photon has to travel twice to make a tick will increase since the photon travels at the finite speed of light.

Length Contraction of Bodies in Motion—Lorentz Contractions

So if time necessarily must be affected by speed, then what else gets affected? If we go back to our example with Sam's coffee shop and our car, what else would change from Sam's frame of reference? After all, previously I said that *all* information from any moving object can only communicate it's attributes at the speed of light. Keeping this in mind, let us explore Sam's frame of reference further. If Sam were looking at us through binoculars, she would notice something peculiar if we were traveling at near light speeds; the length of our car would appear *compressed towards the direction of travel.* In other words the car would be *shorter* than it would be at *rest.* This is explainable if we think about our previous time dilation example with our "simple clock" discrepancies. If time necessarily increases with velocity, and velocity is equal to the distance divided by the time

$$v = \frac{d}{t}$$

then it follows that if we solve for *d* in this equation we get

$$d = vt .$$

Therefore we see that as time *t* increases in this equation, and velocity *v* stays constant (as it is the speed of light), then the distance *d* increases as well—however this cannot be so, for as we know from the negative results of the Michelson-Morley experiment, there are not any observed discrepancies in times regardless of how a photon beam is split up at right angles and then recombined—despite the motion of the earth in its orbit. So if the velocity *v* of light is a constant and if the time *t* of the Michelson-Morley experiment is always equal regardless of the earth's motion, then the only variable in our velocity equation that is left to change is the distance *d*. In order to keep everything else in agreement in this equation as dictated by the Michelson-Morley experiment, then the distance *d* must *decrease* to maintain this equivalence. This is why Sam will see a decrease in the length of the car in the direction of the velocity. This decrease in distance applies to not only how measured distances will be calculated from moving bodies; it will also affect the physical length of an object in the direction of motion as seen by an outside observer as well (by outside—I am referring to an observer outside the reference frame, like Sam).

This amount of length contraction was discovered by Hendrik Lorentz and is called the *Lorentz contraction* (sometimes it is referred to as the *Lorentz-Fitzgerald contraction* as George Fitzgerald contributed to this concept alongside Lorentz). The amount of this contraction is given by the equation:

$$L_{\parallel} = L_o \sqrt{1 - v^2/c^2}$$

Where $L_{||}$ is new length of an object that is traveling *parallel* to the length being measured, L_o is the length of the object at rest, v is the velocity of the object, and c is the speed of light. Once again, for everyday velocities, this contraction in length is negligible, however is nonetheless present. It should also be noted that this contraction in observed length takes place only to lengths that are parallel to the direction of motion, the rest of the object's size attributes will appear unchanged (for instance a circle will appear to contract into a ellipse when at high speeds while maintaining the same diameter perpendicular to the direction of the motion—see below).

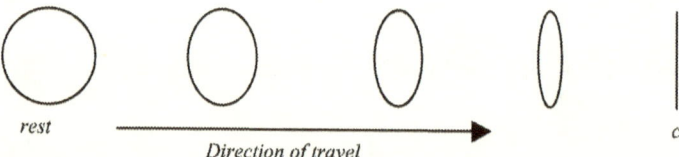

rest   Direction of travel   c

Progression of how a circle would appear from a stationary observer with increasing velocities (from left to right) approaching that of light itself (c), where the circle would appear to have no length <u>in the direction of travel</u>.

Curiously, the Lorentz-Fitzgerald contraction was originally put forth as the presumed consequence of the proposed aether's pressure against moving bodies, thereby contracting them along the direction of their motion (which was proven false as we discussed previously). The figure below compares the length of a 1 km long object moving at increasing velocities as observed by an observer at rest:

Rest Length (km)	Velocity of object (km/sec)	*Observed Length at Velocity (km)
1	10.00	0.99999999944367500000
1	100.00	0.99999994436749600000
1	1,000.00	0.99999443673424500000
1	10,000.00	0.99944352013705400000
1	100,000.00	0.94272742316888100000
1	150,000.00	0.86582546589248300000
1	200,000.00	0.74494293578673000000
1	250,000.00	0.55190009509556000000
1	260,000.00	0.49784391339832000000

1	270,000.00	0.434601091964703000000
1	280,000.00	0.357326679739931000000
1	290,000.00	0.253498131864727000000
1	299,000.00	0.072661705096318600000
1	299,791.00	0.003118766641394880000
1	299,792.45	0.000231020030142532000
1	299,792.46	0.000081677914696989500

as measured from an observer at rest

Mass and Momentum Increases with Velocity

At this point we have seen that time and lengths change with velocity. As we shall explore here, *mass and momentum increase with velocity as well*. Again, this is a direct result of the mathematical "bookkeeping" that needs to take place to ensure the constancy of the velocity of light. There are no exceptions in our modern under-standing of physics that break this rule (in the quantum physics world—we would have to say that the *average* velocity of light is constant, as there are small probabili-ties of photons traveling faster than the speed of light for the briefest instants of time based on the uncertainty principle). What are the consequences of this in relativity? How is this speed limit kept in check by nature? Put differently, what would need to happen to a moving mass to ensure that it could never reach or exceed the speed of light? You increase the *mass with increases in velocity*. As an object achieves more velocity, it also gains more mass that subsequently has more inertia that necessarily requires more *energy* to accelerate it to higher velocities. The formula for the increase in mass with velocity is given by:

$$m = \frac{m_o}{\sqrt{1 - v^2/c^2}}$$

Where *m* is the mass at velocity *v*, and m_o is the rest mass of the object, and *c* is the speed of light. Once again, the affects of this modification to the mass as introduced by Einstein is negligible for objects traveling at speeds much slower than light, allowing our approximations to maintain their integrity for our everyday velocities. The on the following page summarizes a few sample velocities and their associated masses at those velocities according to special relativity as you approach the speed of light:

Rest Mass M (kilograms)	Velocity V (kilometers/second)	Einstein's Mass at Velocity V (kilograms)
1.00	100.0000	1.00000005563251
1.00	1,000.0000	1.00000556329671
1.00	10,000.0000	1.00055678970520
1.00	100,000.0000	1.06075200044420
1.00	200,000.0000	1.34238470084142
1.00	250,000.0000	1.81192213751449
1.00	260,000.0000	2.00866169714504
1.00	270,000.0000	2.30096062455641
1.00	280,000.0000	2.79855957223183
1.00	290,000.0000	3.94480224624939
1.00	295,000.0000	5.61511711316893
1.00	296,000.0000	6.30685450713451
1.00	297,000.0000	7.34371332357197
1.00	298,000.0000	9.15842921312844
1.00	299,000.0000	13.76240756632980
1.00	299,500.0000	22.64485499101560
1.00	299,600.0000	27.91239367431790
1.00	299,700.0000	40.26767860535430
1.00	299,790.0000	246.94788135239500
1.00	299,791.0000	320.63957165860400
1.00	299,792.0000	572.08792402792100
1.00	299,792.4000	1,607.61251344164000

1.00	299,792.4500	4,328.62899110104000
1.00	299,792.4560	8,657.25797319754000
1.00	299,792.4570	12,243.21168959590000
1.00	299,792.4579	38,716.43932490720000
1.00	**299,792.458 = c**	*infinity*

The Equivalence of Mass and Energy

As you may remember, I mentioned earlier that an increase in mass "necessarily" requires more energy to be accelerated to a higher velocity. This is due to the fact that Newton's formula for momentum *p* is given by:

$$p = mv$$

Where *m* is the mass and *v* is the velocity. As you can see, if the mass *m* becomes larger in this equation, then the product of the mass times the velocity will be larger as well. If we include Einstein's modification of mass with velocity, then the formula for momentum *p* becomes:

$$p = \left(\frac{m_o}{\sqrt{1 - v^2/c^2}} \right) \cdot v$$

Where in this case, the mass increases with velocity, the momentum becomes larger and larger as velocity *v* increases, with the whole equation approaching infinity as one approaches *c*, the speed of light.

In our everyday experiences with momentum we intuitively know that a more massive object traveling at high speeds has more energy than a lower mass object traveling at slower speeds. If anyone had to decide between an individual throwing a bowling ball or a pebble at them at equivalent velocities, they would choose the pebble since it has less of an impact, or less *energy* when it hits you (less inertia). However, if a lighter object is traveling fast enough, it can also have a lot of energy for its mass—like a rifle bullet. The point I am getting at here is that *mass and energy are equivalent*. The faster an object is traveling the more its mass increases; the more the mass increases, the more energy it takes to stop, slow-down, or deflect the path of that mass (the inertia increases). A good example of the mass increase and the accompanying energy increase with velocity can be seen in modern day particle accelerators. A synchrotron uses high speed electrons that are traveling very close to the speed of light, and smashes them into other particles to break them apart and observe their mass based on their deflections and trajectories. In order to keep an electron contained to a certain trajectory at these tremendous speeds, the synchro-tron uses powerful magnetic fields that are 2000 times more powerful than New-ton's unaltered laws of physics would have required. This equivalence of mass and energy is the basis for what is perhaps the most famous equation in physics:

$$E = mc^2$$

Where E is the total energy as given by the *mass* of an object times the square of c, the speed of light. As you can see, there is no velocity v in this equation. The only velocity given is the speed of light, which is constant. This means that the total energy E in this equation is the same regardless of how fast the mass m is moving. The figure below illustrates the amount of time a typical light-bulb would be kept lit if all the matter of the masses in the first column was converted entirely into pure energy:

Mass (kg)	Energy (Joules)	*Time this Energy would keep a bulb burning (in years)
0.001	89,875,517,873,682	28,499
0.010	898,755,178,736,818	284,993
0.100	8,987,551,787,368,180	2,849,934
1.000	89,875,517,873,681,800	28,499,340
2.000	179,751,035,747,364,000	56,998,680
3.000	269,626,553,621,045,000	85,498,019
4.000	359,502,071,494,727,000	113,997,359
5.000	449,377,589,368,409,000	142,496,699

6.000	539,253,107,242,091,000	170,996,039
7.000	629,128,625,115,772,000	199,495,378
8.000	719,004,142,989,454,000	227,994,718

A typical 100 Watt light bulb requires about 6000 Joules of energy every minute.

In collisions between quantum objects such as electrons and protons that are produced in high energy accelerators, Einstein's equivalence of energy and mass via $E=mc^2$ plays a crucial role. If two quantum objects collide at high enough energies, they will break apart into a shower of *new* quantum objects. The new objects produced are brand *new* particles that are created out of the energy of the collision since mass and energy are equivalent; they are *not* fragment particles of the impacting "parent" particles. An interesting side effect of this is that you can have collisions between quantum objects where the objects created after the collisions have a greater mass than the objects that were in the initial collision as long as the energy in the collision is larger than the required *rest mass* of the object that got created. If we solve $E=mc^2$ for the mass m, then we have:

$$m = \frac{E}{c^2}.$$

As we can see from this equation, as long as we have enough energy of motion E, we can produce an object of *rest* mass m in the collision. Because of the mass increases associated with velocity in relativity, when two moving objects collide and come together, they must necessarily have greater mass than the two rest masses of the objects. As Michio Kaku states in his book, Parallel Worlds (pg. 33): "If an object becomes heavier the faster it moves, then it means that the energy of motion is being transformed into matter."

Can you catch a Photon?

Based on the information on special relativity that has been presented to this point, we can now tackle a question that intrigued Einstein from an early age: what does a photon of light look like if you were to catch up to it? Einstein imagined that if you could catch up to a photon, it would appear brilliant and stationary, and able to be scooped up like fresh snow. However, with the theory of special relativity he found another answer: you can never catch up to a photon to see what it would look like. As we discovered earlier, *all* frames of reference are equivalent by the principle of relativity. That is, regardless of how fast your frame of reference is traveling, the speed of light is *always* constant—period. If you were to attempt to chase down a photon speeding away from you, no matter how close to the speed of light you got, that photon would still always be speeding away from you at the constant velocity of 186,000 miles per second—no matter what. Regardless of how fast you are going, you are always contained *within* the frame of reference that you are observing the photon from. It is impossible to isolate yourself from your local coordinate system. And, by the principle of relativity, the laws of nature that held true when Einstein discovered them on planet earth, also apply to *anywhere* in the cosmos; including you, traveling at $0.9999c$ trying to catch up to that photon.

This constancy to the velocity of light is something that James Clerk Maxwell noticed when he came up with his four equations unifying the magnetic and electrical forces into electromagnetic waves. The structure of his equations necessitated that the velocity of light is constant *from all perspectives* regardless of who measures it. This notion from Maxwell's equations for electromagnetism goes against Newtonian physics. In Newtonian physics, velocities are *additive*, and there is no limit on the speeds you can achieve. According to Newton, if you are traveling at the speed of light, and turn on a flashlight, then the photons emitted from that flashlight will be traveling at twice the speed of light. Through the genius of Maxwell and Einstein, this concept was effectively shattered (however—as always, we can still use Newton's physics to great accuracy under all but the most extreme velocities).

Generally when we specify velocities, those velocities are expressed according to a point of reference. For instance, a train is said to be traveling 60 miles per hour with respect to the stationary ground. Interestingly, when Maxwell discovered his equations for electromagnetism, and used them to calculate the speed of electromagnetic waves, the answer was the value of the speed of light; however they did not specify what this speed was in reference to. Electromagnetic waves are traveling at the speed of light according to whose perspective? His equations did not provide an explanation for this query; they merely have him an answer. It was Einstein who used Maxwell's calculations on the velocity of electromagnetic waves to hypothesize that there was no reference to base light's velocity on, because *light's velocity is the same for all observers, regardless of their relative motions.*

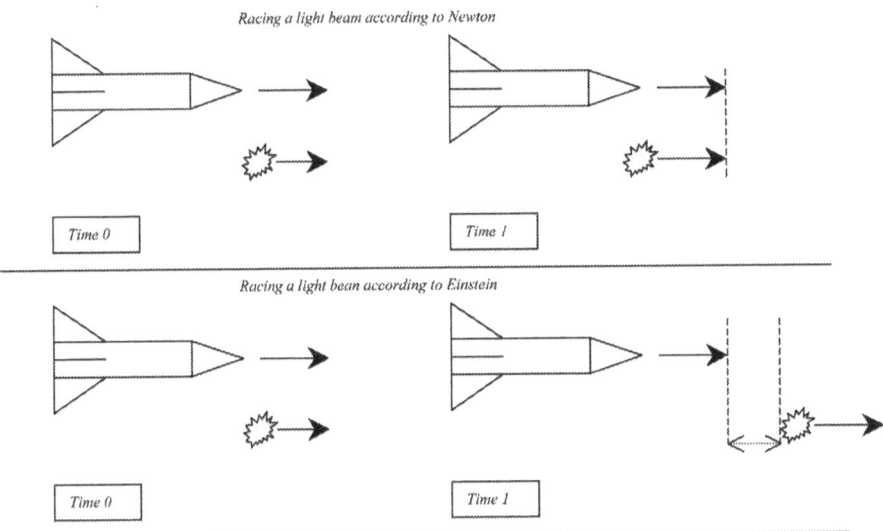

Racing a light beam according to Newton

Time 0   Time 1

Racing a light beam according to Einstein

Time 0   Time 1

Top figure: According to Newton, if you could race a photon of light that started right next to a rocket-ship that could accelerate instantly to the speed of light (which of course is impossible), at "Time 1" in the future, the rocket ship and photon of light would be neck and neck as they are traveling at the same speed.

Bottom figure: According to Einstein, if you re-run the top race under this mathematics, then at "Time 1" the photon will be traveling away from the rocket ship at the speed of light, despite the rocket travel at light speed as well. Einstein showed that you can never catch up to a photon of light, it is impossible; and time and time again, experiments validate this result.

Special Relativity Restrictions—Reiterated

Once again, remember our aforementioned restrictions. First special relativity assumes that all the motions are uniform straight-line motions (no accelerations allowed). Second, special relativity assumes that the object is not rotating as it is moving. Third, special relativity assumes that the conditions of uniform non-rotational velocities are occurring in the absence of a gravitational field.

Einstein's General Relativity

After covering the restrictions associated with the application of special relativity you may be a bit concerned about the practicality of relativity. After all, how often in nature to we encounter non-rotating, constant velocities? Special relativity was discovered by Einstein in 1905. It was not until 1916 that he published his results on General Relativity which greatly expanded the areas of application for special relativity. No longer are the restrictions of uniform motion, the absence of gravitational fields, and no accelerations a concern to us in general relativity. All of the restrictions of special relativity are able to be handled under the umbrella of general relativity. Let us walk through how this came about.

General relativity is an updated theory of Newtonian gravity. This updated theory utilized the concept of space-time, showing that time is no longer an absolute, as Newton thought. In Einstein's relativity, everyone has their own notion of time. People "slice" up space-time at different angles according to their individual frames of reference. Prior to Einstein's usage of another dimension via space-time in general relativity, one of his former professors, Hermann Minkowski, had re-written Einstein's equations for special relativity into a space-time framework. Einstein originally did not like the idea of using four dimensions, however after seeing how using the concept of space-time simplified the mathematics, he ended up using it. Not without its humor is the fact that when taking Minkowski's classes, Einstein was known for missing lectures as he did not see the point in such abstract ideas as extra dimensions. Due to his lack of interest in his subjects, Minkowski's often quoted concept of Einstein at the time is that he was "a lazy dog". There were many other mathematicians that understood the concept and mathematics of four dimensional spaces more so than Einstein. Despite this, it was still Einstein that put forth the effort to see four dimensional space-time to the fruition of general relativity. After the completion of general relativity in 1916, Einstein was in extremely poor health due his obsessive work habits. Severe digestive problems, a gallstone, and an ulcer made him completely bedridden from the winter of 1917 through the spring of 1918.

The Equivalence of Gravity & Constant Accelerations (Inertia)

In order for the principle of relativity to be expanded so that all the laws of nature were equivalent in different reference frames, Einstein needed to show that gravity could be included; however there were problems with this. We can shed some light on the issues of gravity with an everyday example. Let us imagine ourselves at a local shooting range, where we bring a powerful rifle. However this is no ordinary shooting range, as there is a switch inside the range where you can turn gravity on or off when you are shooting. Additionally, the shooting range is immense—several football fields in length, allowing for practicing marksmanship over extraordinary distances. Let us now take out our rifle and start firing at targets at different distances. At close distances, we notice that we can sight in on the center of the target and hit it *nearly* dead-on. However, as the distances become greater and greater, we notice that we cannot simply aim at the middle of the target. We find that when we fire directly at the center of the target at longer distances, we need to aim slightly *higher* than the center of the target to hit it. This is because gravity is *pulling* the bullet down as it is traveling over a distance, thereby necessitating that you aim higher and higher above the center of the target as the distances become greater and greater. When we turn the gravity off at these longer distances, we notice that the bullets go in perfectly straight lines and we can go back to aiming directly at the center of the targets, just as when gravity was on and were very close to our targets.

Now let us consider that someone magically beamed our shooting range into the vacuum of space, away from all appreciable gravitational sources, and accelerated the whole shooting range at a constant acceleration that is equal to that of gravity on earth. If we start shooting again, we would notice the exact same results as we did on earth, when we were shooting with the gravity "on". The key point here, which is what Einstein realized, is that *we would not be able to distinguish experimentally the different between the effects of gravity and the effects of a uniform acceleration.* Or to expand on this, *there is no way for us to tell the difference between being at rest on earth shooting our rifle, and shooting our rifle in an accelerating reference frame that is in space.* Likewise if you are falling at just the right velocity inside a gravitational field, you will feel weightless just as if you were in space. If you were in a room that was free falling to the earth with no windows and no air resistance, eventually your acceleration would match that of gravity and you and your frame of reference in the falling room would be increasing velocities at the same rate and you would feel weightless. If you took the same windowless room and placed it in space, it would feel the same. Moreover, if you were a scientist taking measurements from a windowless laboratory, you would not be able to tell the difference between being in a uniform gravitational field, and being accelerated at an equivalent uniform rate. Additionally, if you were weightless in your laboratory, you would not be able to tell if you were free-falling, or were in the vacuum of space. This means that *an accelerating frame of reference is equivalent to a system that is at rest in a gravitational field* (this is called the *principle of equivalence). You* only feel the effects of a gravitational field when you *resist* its acceleration. In the absence of resistance in a gravitational field it is equivalent to being weightless in space. It is important to interject a key point at this juncture. The principle of equivalence requires the above events to be occurring in a vacuum, where any outside effects such as air resistance, would be eliminated.

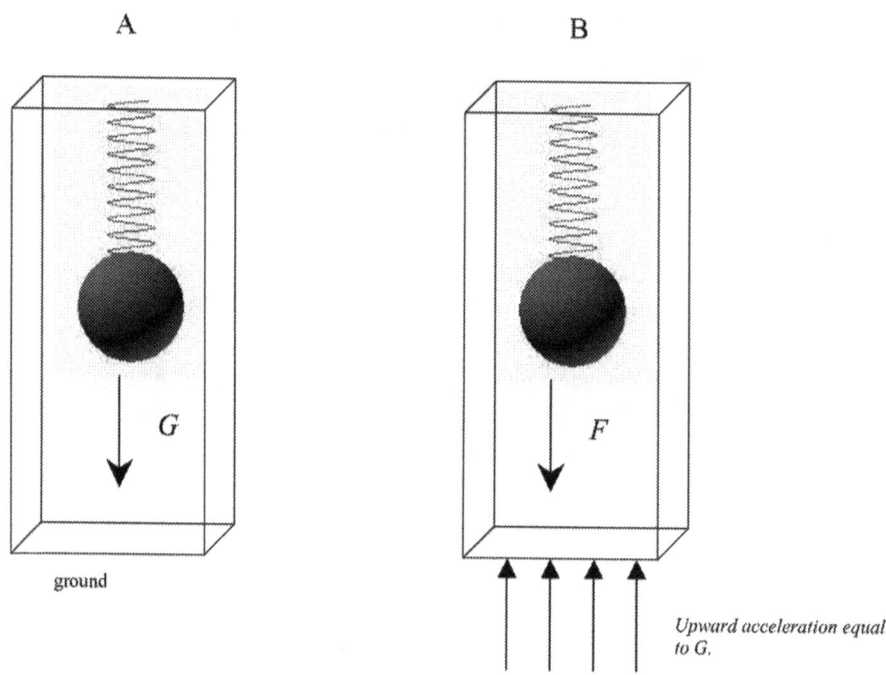

A   B

G

ground

F

Upward acceleration equal to G.

Elevator A and B are both equipped with identical force meters contained within the spheres. The spring measures the amount of downward force that is present. Elevator A is at rest on the earth, with its gravitational field G pulling downward. Elevator B is in outer space, however it has thrusters that accelerate it upward at a rate that is equal to G. The principle of equivalence states that there is no way to tell the difference which elevator you are inside of (if there are no windows—of course). Gravity and acceleration are equivalent.

Clocks in Gravitational Fields

Einstein showed that clocks in gravitational fields behave in the same way that clocks did in the equations for special relativity. We can show this by example. Let us imagine this time a huge rocket that is in constant acceleration through the vacuum of space. If we placed a clock at the head of the rocket and a clock at the rear of the rocket, with each clock sending out a photon of light after each second that has passed, what would we observe? Well, if we are sitting at the rear of the rocket, the clock at the front of the rocket would appear to be running *faster* than the clock that is next to us at the rear. The reason for this makes sense after the application of some critical thinking. After a second goes by the front clock emits a photon traveling at a *finite* speed, the speed of light. So, necessarily, it takes a certain amount of *time* for the photon to reach us at the back of the rocket. However, as we know, the rocket is accelerating as well, so for every increment of time that the photon takes to travel to the rear of the rocket, the rocket is accelerating the rear of the rocket *towards* the incoming photon, thereby *decreasing* the distance it needs to travel. This translates into a faster running clock.

If we now take our space ship and place it in the gravitational on the earth's surface, what would we see? By applying the principle of equivalence to this experiment, we must conclude that the laws of nature must hold true for a uniformly accelerating frame of reference, and a frame of reference that is a rest in a uniform gravitational field. This means a clock in a stationary vertical rocket ship that is at rest on the surface of the earth would exhibit the same behavior as our previous example in space. Therefore, a clock that is higher off the surface of the earth will run faster than a clock that is on the ground. This has in fact been measured experimentally to a high degree of accuracy that is in accordance to Einstein's predictions in general relativity. (It should be noted that in fact the earth's gravitational field is not perfectly uniform. Since the earth is circular, the gravitational field lines must *converge* at the center of mass [just as a magnets field lines converge on the poles of the magnet], in the earth's center. However for the short distances used in our rocket example, such effects can be ignored without detriment). These gravitational effects in timekeeping in relation to height are extraordinarily small. In the earth's gravitational field for instance, the difference in time between a clock on the ground and a clock on top of a tall building is less than two parts in a *million-million*.

This concept of dynamic time based on frames of reference and their associated velocities is a striking departure from Newton's views of time. According to Newton, time was absolute for the universe; a concept that he referred to as *absolute time*. To Newton, a watch at rest kept exactly the same time as a watch strapped to an object traveling at the speed of light. In special relativity, there is no such thing as absolute time, however there is *absolute space-time*. In other words, the universe *itself* which is housed in Einstein's space-time fabric is a frame of reference from which we are unable to isolate ourselves, thereby making it *absolute*. No matter what event occurs in our universe, it is transpiring within the space-time of the universe itself.

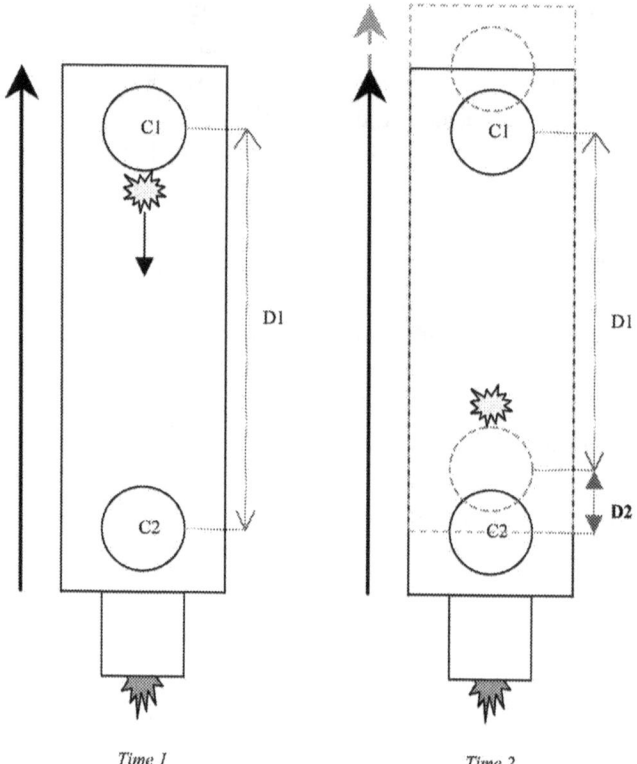

Time 1   Time 2

Both diagrams represent a box in space that has a rocket engine attached at the bottom that provide an upwards acceleration. There is a clock C1 at the top of the box and another clock C2 that is at the bottom of the box as it is accelerating. At constant increments, C1 will emit a photon of light towards C2. Time 1 represents the instant that such a photon is being emitted by C1. At Time 2, the box has moved upwards in the finite time that it takes the photon to travel to C2. This upward movement by the box, shortened the distance D1 that the photon has to travel to be recorded at C2 by the amount D2. This means that according to C2 at the base of the box, C1 is running faster than normal. From Einstein's principle of equivalence, the same must be true for an object at rest in a gravitational field, since an accelerating reference frame and a reference frame at rest in a gravitational field of the same magnitude as the acceleration are equivalent. This effect on earth has been proven experimentally, confirming Einstein's insight.

The Curvature of Space-Time

According to Newton's first law of motion, a body will continue to travel in a *straight line* unless acted upon by an outside force. The key here is that in the presence of gravitational fields or accelerated motions, Newton's traditional straight line becomes a *curved line*. This means that gravity is nothing more than the curvature of the space-time around it. The more mass an object has, the more it curves the space-time around it, just as a marble sitting on a taut rubber sheet will stretch the rubber very slightly as compared to a bowling ball. In order for Einstein to include gravity into general relativity, he had to geometrically alter his equations. In special relativity, the equations were based on Euclidean geometry, what we all used in high school math class. In Euclidean systems, there are the obvious features that we are used to, such as the interior angles of triangles always adding up to 180 degrees, lines being perfectly straight, the angles of a square adding up to 360 degrees, the circumference of a circle divided by its diameter is *pi* (3.14159265…)…etcetera. The problem with using Euclidean in relativity is readily apparent around us: there are not any truly straight lines in nature. When you throw a ball, it follows a curved path. When you fire a bullet, if follows a curved path. To solve this problem, Einstein left the Euclidean geometry behind, and replaced it with a new geometrical representation called *Gaussian geometry*. This geometry is sometimes referred to as *Riemannian geometry* as well after Bernhard Riemann. In a more precise sense, non-Euclidean geometry was a combination of efforts by Riemann, Gauss, Bolyai and Lobachevsky. The best way to think of Gaussian geometry is to think of traveling around on the *surface* of a giant sphere, such as our earth (in actuality the earth is not a perfect sphere; however it still works for our purposes). The key difference between Euclidean and Gaussian geometries is that Euclidean geometry is based on projecting geometrical figures onto *planes* while Gaussian is based on projections onto *spheres*. This new geometrical system is exactly what Einstein needed for the equations of special relativity to take into account the curved trajectories of objects traveling in gravitational fields.

There are some interesting differences between Euclidean and Gaussian geometries that we can profit from exploring. Let us start with distances. In the world of Euclidian geometry, the shortest distance between any two points is a straight line. For Gaussian geometry, where we must picture ourselves on the surface of a sphere, the shortest distance between points is a *curved* line called a *geodesic*. When you fly in an airplane between two countries on a map, the map may be projected in a Euclidean manor on the 2D plane of the map's paper surface, however in reality the airplane is traveling in a curved path to match the curvature of the earth below (if you did travel in a straight line, the airplane would eventually end up in outer space). Moreover, the farther you travel on a sphere, the more overall curvature your path will have. For instance, if you walked from Omaha Nebraska to Kansas City Kansas, you will have traveled such a short distance compared to circumference of the earth (about 25,000 miles), that your path is nearly linear and Euclidean. However, if you walked from Omaha to the South Pole, you have added considerable distance and hence curvature to your path. If you walked long enough around a sphere in any direction you could arrive back at your starting point and your path would be curved a full 360 degrees from walking in what you thought was a "straight" line.

This brings us to another peculiar feature of Gaussian geometry: circles. In Euclid geometry, the famous ratio *pi* is derived from taking the circumference of a circle divided by diameter (or 2 times the radius—2*r*).

$$\pi = \frac{circumference}{diameter} = \frac{c}{d} = \frac{c}{2r} = 3.14159265...$$

This value is constant for all sized circles in Euclid geometry. In Gaussian geometrical space, if we made the same circle, we would notice that *pi* was no longer a constant. If we make a suitably large diameter circle on the surface of the earth and then plug the numbers into our equation above to find the ratio between the circumference and the diameter, we would find that it is always *greater* than the value of pi above. If we kept repeating this process for larger and larger circles, we would notice larger and larger values for the ratio of circumference to the diameter. Just as we saw that greater distances between two points on a sphere equate to *greater* curvature, the same rule applies to circles projected onto a sphere as well. As it turns out, the curved projection of Gaussian coordinate systems creates an *excess or reduction* in the summation of angles, areas and volumes as compared to equivalent Euclidean objects. For example, a triangle in Euclid coordinates will always have interior angles that sum to 180 degrees. In a Gaussian coordinate system, the angles of a triangle will always sum to a value *greater or less* than 180 degrees and *the amount of this difference will increase as the size of the object increases, or as the amount of curvature increases*. This is true for any comparison of two dimensional and three dimensional objects in Euclidean and Gaussian coordinate systems. This is an important concept as it provides a way to establish the *average* curvature in an area or volume of space-time since the differences in the Euclid representation of the area or volume compared to the Gaussian representation of the area or volume are proportional to the curvature in that area or volume. A more precise way to calculate the average curvature by means of the excess radius is given by:

$$R_{excess} = \frac{G}{3c^2} \cdot M$$

Where R_{excess} is the excess radius, G is Newton's gravitational constant (6.67 x 10^{-11} Nm^{2}kg^{-2}), c is the speed of light, and M is the mass. If we plug in numbers for the mass of the earth, we find that roughly speaking it has 1.48mm more radius than it should have for its surface area. Plugging the numbers in for our sun, we find that it has 0.492 km more radius than it should have for its surface area. This type of relationship to the excess in the radius of a sphere for its surface area being proportional to the mass contained within it is of prime value to astronomers and cosmologists. This allows them to estimate the curvature of the entire observable universe, and provides insight as to whether the cosmos is, on average, flat like a plane (an "open" shape), or round like a sphere (a "closed" shape) There are also possibilities for *negative* curvatures where the sum of the angles of a triangle would add up to *less* than 180 degrees.

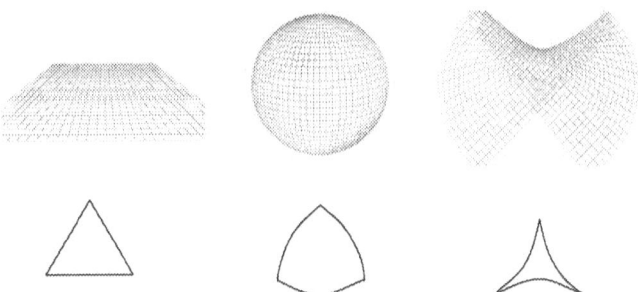

Properties of triangles on different curved surfaces. On the left, a regular triangle on a flat surface will have angles that sum up to 180 degrees. On a positively curved object, such as a sphere, a triangle will have angles that sum up to more than 180 degrees as represented by the middle pictures. On the right is a triangle on a negatively curved space, such as a saddle, where the sum of the angles will sum up to less than 180 degrees. Knowing these types of geometric features of curved spaces and their effects on geometry allows us to calculate the curvature of space-time as proposed by general relativity.

In Einstein's general relativity, the degree of curvature in space-time is proportional to the amount of matter present in that volume. The more matter you have in an area, the greater the curvature of space-time in that area. According to general relativity, gravity is nothing more than a geometrical distortion of space-time that is introduced in the presence of matter. When the amount of matter is on the order of our everyday experiences, the equations of general relativity can be reduced down to Newton's laws without any problems. However, as masses increase to larger values, we must use general relativity to remain accurate in our portrayal of nature, and we must leave Newton's laws behind. While Einstein was certainly the first person to fully work out the details of using geometric distortions to explain gravity, he was not the first to think along these lines. William Kingdon Clifford, in 1870, published a paper in which he explored the plausibility of thinking of matter in terms of bumps in space and that the known forces of magnetism, electricity, and *gravity* could be explained by changes in the *geometry* of space. This line of thinking is also used to various degrees in even more modern theories such as superstring theory and M-theory. Many great thinkers that were well ahead of their time and were labeled as scientific or mathematical "misfits" have ended up making important contributions to physics and math long after their passing, when our knowledge base has finally caught up to their visions.

Originally, when Einstein was exploring the thought of adding an extra dimension into general relativity to represent the non-Euclidean coordinate systems that would be needed, he found himself without the mathematical knowledge of higher dimensional spaces (called *hyperspaces*). In one of the few times that Einstein asked for assistance when engrossed in the challenges of formulating his general theory of relativity, he implored the assistance of his good friend and mathematician Marcel Grossman, to get him up to speed on the mathematical concepts needed to formulate relativity in four dimensional space.

Proof of Curvature of Space-Time

Over the decades, general relativity has withstood the scrutiny of many tests. While general relativity states that *any* mass or energy will distort space-time to some degree, it requires an extraordinarily large mass to make these curvatures of space-time large enough for them to be tested experimentally. For small masses, the space-time distortions are so minor as to allow a disregard of their presence, as we do in our daily experiences. So how can you test experimentally whether relativity is correct in its prediction of space-time curvature? Einstein proposed a novel validity test: use the glare-masking properties of a total solar eclipse to enable the precise measurement of the positions of stars that are nearly grazing the limb of the darkened sun where the geometrical distortions of space-time would be most significant, and then measure the difference in their positions due to the bending of their light rays from the curvature of space-time around the sun, to their locations in sun's absence. A deviation in the positions of stars near the sun before and after a solar eclipse could provide explicit proof for general relativity. Einstein predicted that such a mass as the sun should be able to bend the rays of light from a distant star far enough to be measured experimentally. Einstein calculated that stars rays that were just grazing the surface of the sun would be deflected by 1.7 seconds of arc. Later expeditions to measure this bending of starlight came back a resounding success—agreeing with Einstein's predicted number to a high degree of accuracy. This successful validation of general relativity made Einstein an overnight celebrity, vaulting him to his position of being the most popular physicist in history.

This bending of light is often referred to by astronomers as *gravitational lensing*. On cosmic scales, astronomers have obtained many pictures of this effect where light emitted from distant galaxies is bent into arcs or rings (called *Einstein rings*) from a conglomeration of mass between earth and the distant galaxy (such as other galaxies or large clouds of gas). This lensing effect can also amplify the light of a distant galaxy by focusing it, like a magnifying glass, thereby increasing its apparent brightness from here on earth. Also, astronomers sometimes see what appear to be mirror "copies" of distant objects that are believed to be caused from a single distant object's light being split apart, creating illusions of more objects, as the distant photons get bent by space-time curvatures caused by intervening matter in our line of sight.

The Concept of Space-Time

It is worth expanding for a short time to explain Einstein's concept of *space-time*. If we remember that an object's dimensionality is simply a measure of the unique degrees of freedom it possesses, then space-time becomes an understandable concept. As we covered previously, if you have an important meeting in a downtown office building, you need four pieces of information to get there: you need the x and y coordinates as provided by the street address, you need the z coordinate of the floor number, and you need to know what *time* the meeting is occurring.

While this example is not a mirror-image of true space-time in Einstein's line of thought, it still gets the point across. Einstein made it clear that if a mass located at a point $p(x,y,z)$ is curving its surrounding space geometry, thereby affecting the motions of bodies near $p(x,y,z)$, and since time is nothing but motion, then time is defined differently in and near that point $p(x,y,z)$. Therefore time also must be included in every points definition—$p(x,y,z,t)$. This is a more accurate way to envision space-time. Since time is motion, and motion is determined by the bending of space, then the two must necessarily be woven into a single fabric—that fabric is space-time. In our everyday lives where the velocities of events are occurring at such slow speed as compared to what is needed for relativistic effects to take place, Einstein's and Newton's equations of gravity become, for all practical purposes, equivalent. We don't need to worry about time dilation, mass increases, or length contractions on our way to meetings downtown.

Gravity's speed limit

Up to this point, we have been repeatedly hammering in the concept of nothing being able to travel faster than the speed of light. What about gravity? What is the speed at which gravity is able to propagate? In Newton's non-relativistic view, gravity was an instantaneous force regardless of the distances involved. In Newton's mind, gravity treated all objects as if they were all stacked on top of each other with no distance separating them. However, Einstein found through his general relativity theory that gravity is bound by the speed of light as well. The speed of light is an absolute speed limit—it does not matter what type of force you are dealing with, be it the electromagnetic, weak, strong, or gravitational forces. The classic example of this is to think of the moon in orbit around the earth. According to Newton, if the moon were to suddenly vanish while you were standing on a beach watching it during a high tide, the water would *immediately* start receding. However, the corrected view of this as put forth by Einstein's general relativity is if the moon suddenly vanished, it would take a certain amount of time for the water to "know" that the moon was gone. This delay would be exactly equal to the time it would take to travel the distance from the moon to the earth at the speed of light. Newton's idea of "forces at a distance" was incorrect, where objects under the influence of gravity could be imagined as connected by invisible strings that would allow gravitational forces to be instantly carried from one object to another, regardless of separation distance. Gravity is bound by the speed of light as well.

Gravity Waves

Yet another intriguing prediction that was spawned by general relativity is the concept of *gravity waves*. As we covered earlier, gravity is bound by the speed of light. Additionally, we know that according to Einstein's equations, gravity is nothing more than the curving of space-time surrounding bodies with mass. According to general relativity, a gravity wave is produced whenever any object with mass changes direction in space-time. Just as a boat moving through water generates a "wake" that spreads across the water as it moves, so too does a mass traveling through the fabric of space-time generate a "wake" of gravity as it moves, in the form of gravity waves. While no gravity waves have been detected experimentally, they are likely to exist based on the excellent agreement to experiment that other aspects of relativity have been verified. We are at a cusp in experimental capability where gravity waves will hopefully be found in the near future.

In our previous example with the moon vanishing suddenly during high tide, we can now imagine a gravity wave speeding towards earth at the moment the moon vanished. If we think about this deeper, it makes sense. Let us imagine the moon as a giant ball that is sitting in a puddle of water. It has just enough buoyancy so that half of it is under water, and the other half is above water. If you suddenly make the giant ball vanish by a magical means where we don't have to touch the ball, thereby not disturbing the water, what happens? As soon as the ball vanishes, the volume of space that the ball took up in the water is now empty, and the water rushes in to fill in the space. As it does so, it generates a series of waves as it oscillates up and down in the ball's old location, sending those waves out in all directions. Eventually with enough time elapsed, the water will become calm again. This is analogous to our moon example, except instead of water, we have space-time. In our water example, when we make the ball vanish, the space it used to occupy gets filled in at a velocity that is based on the intrinsic properties of water. In space-time, when objects with mass move, space-time rushes in to fill in where the object used to be with a velocity based on it's properties as well—and Einstein showed through relativity theory that space-time's intrinsic properties make that speed the velocity of light.

For those interested in further exploration of gravity waves, black holes, and general relativity and its implications, there is an overview by the University of Illinois at http://archive.ncsa.uiuc.edu/Cyberia/NumRel/GenRelativity.html. I highly recommend exploring this web site. It contains informative animation sequences with sound, video sequences with insights from professors, and remarkable graphics. The text is highly readable and well organized. For those seeking information specific to gravity waves, you will profit from exploring Caltech's website http://www.ligo.caltech.edu/ for the Laser Interferometer Gravitational wave Observatory, or LIGO. This facility uses the same experimental premise as the Michelson-Morley experiment, where laser beams are split apart at 90 degree angles, bounced back and forth several times, and then recombined in hopes that any slight changes in lengths due to the passing of gravity waves will cause changes in the interference patterns of the recombined beam.

An interesting aspect of gravity waves as predicted by general relativity is that they will lose energy the farther that they travel from their sources. However the information contained within the structure of the gravity waves itself will remain *unchanged* regardless of what sort of matter is in its path. This is a profound implication, for it means that if we can detect and study gravity waves experimentally, then we are receiving unaltered data directly from the source of the gravity wave (e.g. a black hole, supernovae, pulsars…etc). Moreover, calculations show that events that generate gravity waves may be unique to that particular event, thereby allowing scientists to directly calculate the size and mass of the object producing the disturbance. Computer models suggest that with black holes in particular, the gravity wave produced will be unique to all black holes, just as a bell has a unique frequency with which it "rings". With this gravitational wave signature from black holes researchers can calculated the size and rate of rotation of the black hole. In reality, in order for gravitational waves to be produced, certain types of criteria must be met. More specifically, calculations with general relativity tell us that spherical or uniformly rotating objects will *not* produce gravity waves.

As I mentioned earlier in this book, any reader interested in searching for gravitational waves from pulsars can visit http://einstein.phys.uwm.edu and download the free Einstein@home software. Einstein@home enables your computer to search for gravitational waves during your computer's idle time. My own personal experience with this software has found it to be robust and relatively nonintrusive on computing resources.

Further Evidence of the Validity of Relativity

I mentioned the accuracy of Einstein's equations in predicting the amount of deflection that star light grazing the limb of the sun would produce; there is however, more evidence of his theory that we can explored. One of these involves gravity waves and neutron star spin rates. Relativity predicts that spinning, non-spherical objects will emit gravity waves. Neutron stars are incredibly dense stellar cores whose further collapse into black holes is only halted by the repulsive forces between neutrons. As the parent of a neutron star collapses any angular rotations present will be increased as the radius of the compressing core decreases (just as an ice skater pulls their arms in to spin faster). By the time a massive star has compressed down to the size of a neutron star, it can be spinning several times a *second*. This rapid rotation prevents it from being perfectly spherical and they therefore should emit gravity waves. The key here is that any object that emits radiation, even gravitational radiation, will lose energy as a result of that emission. In the case of gravity waves, it is predicted that as a spinning neutron star (a pulsar) emits gravity waves, its rotation rate will decrease slowly, until eventually it stops spinning (and emitting gravitational radiation in the process). Observations of certain spinning neutron stars over time have shown slowing in their rotational rates that are in agreement with the predicted values from general relativity.

Mercury's Orbit

In Newton's theory of gravity, there is no bending of space-time (there is no such thing as space-time in Newton's construct). Mercury is close enough to the sun that the bending of space-time from the sun's mass is enough to markedly change the orbital patterns of Mercury over time. More specifically, if we plotted out a dot for every time that Mercury reached the farthest distance from the sun its orbit, we would notice that over time this series of dots changes slightly with every orbit. In a physicist's world, this is called *precession*. Newton's theory of gravity was able to account for some of this precession, but without taking into account the bending of space-time due to Mercury's proximity to the sun, it never solves the problem precisely. If we apply Einstein's equations of relativity to Mercury's orbit, we find an exact match between the predicted and actual magnitude of precession.

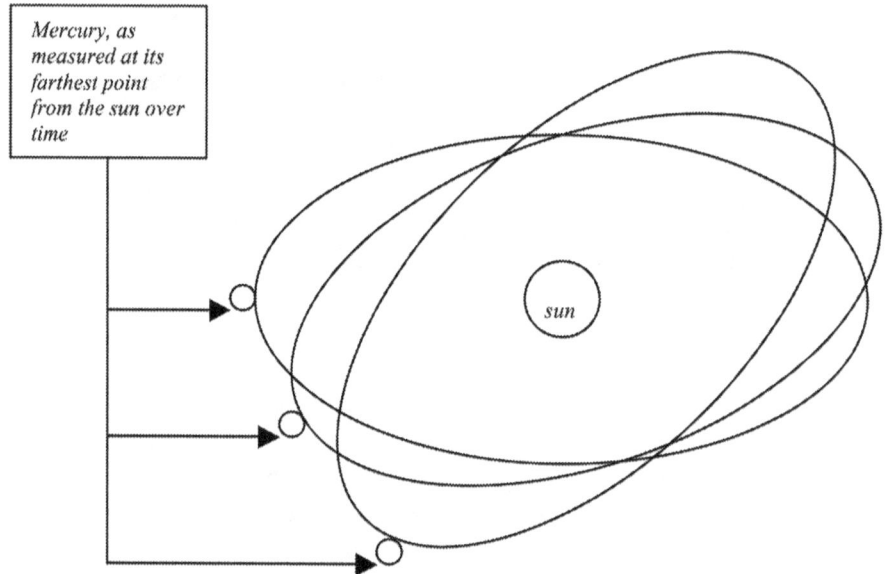

Mercury, as
measured at its
farthest point
from the sun over
time

sun

The precession of the orbit of Mercury over time. When Mercury reaches its
farthest point from the sun in its orbit, the position shifts from the previous
orbit—this is precession. Mercury never traces out the same orbit. Each orbit
around the sun brings it into a slightly different orbit due to its precession.

Gravitational Redshifts—Photons Grappling with Gravity

General relativity predicts that in the presence of a gravitational field, a photon
trying to escape from it will lose energy to the gravitational field. This energy loss
can be seen in the photon's frequency, since as its energy decreases, so does its frequency (as frequency decreases, wavelength *increases*). This effect has been
observed and measured experimentally on earth from high energy photons that
were shot out of the earth's gravitational field, losing energy during their escape,
which lengthened their wavelength by miniscule amounts (which would correspond to losing energy during their escape). The measured reduction in energy
was in tight agreement with the predictions of Einstein's relativity. This energy
loss or lengthening of photon wavelengths in gravitational fields is called *gravitational redshift* (the "red" in "redshift" comes from the visible spectrum, where red
is housed at the lower energy, and hence longer wavelength, part of the spectrum). This redshift has also been measured and confirmed with theory for pho-

tons emitted by the sun which have to escape from its gravitational field on their way to detectors on earth.

Global Positioning Systems (GPS)

Global Positioning Systems rely on a series of satellites orbiting the earth, which transmit their location information back down to earth, where it is received by the ground-based GPS units. The ground based GPS units then lock on to three or more of these satellite signals to then triangulate your position on the surface of the earth. The technology behind GPS requires incredible accuracy in the signals that are being sent down to earth by the satellites—on the order of billionths of a second. In order to maintain the required accuracies to make GPS technology work, the shifts in frequencies due to the satellites velocity in orbit according to Einstein's theory of special relativity must be taken into account. If these slight corrects were no taken into account, and Newton's laws were used for the calculations, then the satellites clocks would run slightly faster each day, eroding the accuracy of the GPS measurements.

7

Quantum Physics

There is probably nothing more confusing than quantum physics. I have often read that if someone claims to have a clear understanding of quantum physics, then they are openly acknowledging that they do *not* understand it. Practically every aspect of this field is counter intuitive to our everyday thinking. Books on quantum physics openly state that there are many aspects of the theory that are unexplainable, and simply need to be taken as a part of the theory—with no reason as to why. As Richard Feynman states in his book QED (Quantum Electrodynamics—Page 10), "...while I am describing to you *how* Nature works, you won't understand *why* Nature works that way. But you see, nobody understands that. I can't explain why Nature behaves in this peculiar way....The theory of quantum electrodynamics describes Nature as absurd from the point of view of common sense. And it agrees fully with experiment. So I hope you can accept Nature as She is—absurd." This is a hard standpoint for many people—just ask any child. There is always the urge to ask "why". On many levels, people do not want to hear the answer "We do not know why, that is just how it is", especially from physicists. However, that is the current reality of quantum physics.

So why would physicists waste any time on a theory that is so "absurd"? The reason is that quantum physics has been testing more thoroughly and its predictions have matched up with experiments more accurately than any other theory ever put forth in physics. Again, to reference from Feynman's QED (page 7): "The theory of quantum electrodynamics has now lasted more than fifty years, and has been tested more and more accurately over a wider and wider range of conditions. At the present time I can proudly say that there is *no significant difference* between experiment and theory!...To give you a feeling for the accuracy...it comes out something like this: If you were to measure the distance form Los Angeles to New York to this accuracy, it would be exact to the thickness of a human hair...Things have been checked at distance scales that range from one hundred times the size of the earth down to one-hundredth the size of an atomic nucleus." Clearly for any theory this is an impressive statement. As Feynman says quantum electrodynamics is the "jewel of physics—our proudest possession." Quantum physics is indeed powerful in its ability to describe many areas of science.

Infinite Complexity

Things on the quantum level are incredible small. They are many orders of magnitude smaller than anything that we see in our everyday experiences. To reach the quantum level we need to start looking at things that occur on scales of *nanometers*. A nanometer is a *billionth* of a meter—a remarkably small scale that is not easily imagined. At such tiny scales complexity abounds, and some of Feynman's "absurdity" must be allowed to creep in. Seemingly simple everyday objects turn into a nearly infinite tapestry of subatomic particles, atoms, and molecules. If you take the edge of any solid everyday object, like a penny for example, and magnify it until you are at the quantum level, it would be a monster of mind boggling complexity with trillions of rapidly vibrating probability shells of atoms (I will explain those probability shells later). For our penny, if we were to try and model it in a modern computer in the most real sense available, taking into account everything that is happening to every atom (and each one of those atoms interactions with its neighboring atoms), and model it through a slice of time—it would be impossible. We would be making calculations for trillions of atoms that need to be individually tracked in our computer for every instant of time. In such a scenario, we would run out of computing power very quickly. Nature houses a profound complexity, even for the simplest of objects.

Planck's Constant, Planck Length, and Planck Time

Planck's constant is the fundamental constant of quantum physics. It is displayed as an h in equations, and is the constant of proportionality in quantum physics—just as G is the constant of proportionality in Newton's theory of gravity. Einstein brought forth the discovery of the Planck's constant by means of discovering that light was the fastest velocity possible—period. This placed a limit on the smallest sizes that can have any meaning to us based on this velocity cap. This equates to a distance of about 1.6×10^{-35} meters—an incredibly small size. This size is what is referred to as the *Planck Length*. The Planck length is best thought of as the length scale when Einstein's classical equations of relativity break down and the quantum effects become significant. The Planck length is the smallest size possible based on the limitations imposed by the speed of light—it is the quantum limit to length (or the quanta of length). The time it takes to travel a Planck length at the speed of light is therefore the smallest amount of time that can have real meaning. This is called the *Planck time*, and is equal to 10^{-43} seconds—this is likewise the quantum limit of time (the quanta of time). Based on our current understanding of the laws of physics, we cannot obtain any information mathematically or experimentally from before 10^{-43} seconds after the Big Bang at the beginning of the cosmos. Before this Planck time, we do not know what happened to our cosmos (in other words according to quantum theory the universe came into existence when it was 10^{-43} seconds old). It is hoped that with time, superstring theory and m-theory will provide a means of probing back in time prior to the Planck time. These theories will be explored later in this book.

The Salient Observer

In quantum physics, it is impossible for the observer to remove themselves from the experiment they are conducting. No matter how the experiment is devised, no matter how clever and careful, the fact that there is an observer that needs to make some sort of *measurement* will affect the outcome of the experiment. The objects being studied on the quantum level are so small, that any measurement will *in itself* cause an infringement upon the otherwise unimpeded results if the experiment was left to run unobserved (and hence unmeasured). Typically the clarification of this concept involves an outline and setup of the famous double-slit experiment, where photons of light are allowed to pass through two extremely narrow parallel slits and then their final positions registered onto a screen at some distance behind the slits. While this double-slit experiment example is invaluable, and will be covered more in depth later, we can achieve the same clarification of the concept of the "observer" and their role in quantum physics by another path.

Light contains energy based on its wavelength. The longer the wavelength, the less energy photons of light have (e.g. radio waves, micro waves). The shorter the wavelength, the more energy photons have (e.g. ultraviolet rays, x-rays, gamma rays). In order to measure the position or velocity of a quantum object such as an electron or photon, we *have* to send another photon towards it to find out that information, much as a police officer uses a radar gun to fire radar wavelength photons at your car to determine your cars position and speed. The problem is that when that police officer is bombarding your car with those trillions of radar photons, they are all smashing into your car and robbing it of a tiny bit of its energy (more specifically momentum) in the process. This may seem like a silly statement, however it is truth. In essence what is happening is that in order to find out how fast your car is going, the police officer's radar gun is taking away a tiny bit of your cars forward velocity from all of those trillions of radar photon impacts to receive the validation they need to give you a speeding ticket. Of course, this deviation is incredible small, and will not give you warrant to protest your heavy foot in court. This velocity "robbery" is unimaginably miniscule since the energy of your moving car that weights thousands of pounds is exponentially more than the energy (which is equivalent to mass through $E=mc^2$) of all those trillions of radar photons colliding with it.

Another comparable example would be to shoot a b-b gun at a bowling ball that is rolling towards you. The b-b may be traveling many times faster than the bowling ball; nonetheless the net effect is practically nil. The energy of the massive bowling ball is too large for the b-b to noticeably slow it down (although it does, a tiny bit). Even the results of this example are *trillions* of times more intense than the effects of our radar photons colliding with your speeding car.

So where are we going with these examples, and what do they have to do with the effects of observations on quantum events? Let us take our police car with its radar gun and shrink it down to the quantum level and have it look for photons that are exceeding the speed of light (upon accepting this assignment, the officer was incessantly ridiculed by his seniors). If we continue to let our police officer use their radar gun, then swarms of photons will pass undetected since the wavelength of the radar photons is huge compared to the diameter of the photons we are trying to catch speeding. The shorter the wavelength we use, the more detailed information about positions, and hence velocities we can achieve. So, let us have our atomic police officer use an ultraviolet gun which utilizes a short-enough wavelengths to detect the velocities of all the photons passing by.

The above sequence simulates the effects of using successively shorter and shorter wavelength light to increase the resolution of the star. The finest details resolvable are directly related to the wavelength of the light being used to study it. The shorter the wavelength of the light, the greater the detail that is visible. The caveat at the quantum level is that longer wavelengths disturb the velocity of the particle being observed less since longer wavelength light has less energy, however this comes at a price, since with the longer wavelength you will give up knowledge of the particles position. Conversely, if you use a high energy photon of short wavelength to study A quantum event, you will have very accurate position information, but the higher energy of the photons will disturb the velocity measurements more. This limit to the precision of measurements is a core concept of the uncertainty principle.

After a short period, a photon comes zipping down the street and our police officer pulls the trigger on this converted ultraviolet-gun. Immediately, and much to the officer's surprise, huge, oscillating bursts of high-energy ultraviolet photons start flying out of the gun towards the suspected speeding photon. What is more, the bursts of energy coming out of the ultraviolet gun are the same size as the entire photon that the police officer is trying to catch speeding. Immediately the officer turns off his gun, however it is too late—one of the giant bursts of energy from the ultraviolet gun has already collided with the photon, hitting it with so much force as to veer it onto another street where it zips out of site.

In our full-scale example, this would be akin to traveling down the street towards a police officer and their radar gun, when suddenly gigantic "bullets" with a size and mass comparable to that of your *entire car* come hurling towards you from the radar gun. Let us say your car has huge bumpers on it that will not let these bullets smash your car. When that radar gun 'bullet" hits your car, *even if it grazes* your car, you will veer off in another direction and velocity. If that police officer was not stationed there with a ridiculously overpowered cannon of a radar gun, then you would have continued along on a completely different path.

This is the problem of observing quantum phenomena. In order to get the details of what is going on with quantum size objects, we must bombard those objects with other quantum sized objects to see what is going on. With this in mind, we must come to the conclusion that if we were not observing the quantum event, then it necessarily would have had a *different* outcome since by observing and measuring the system, we inevitably alter it. We are unable to isolate ourselves from the systems we are studying on the quantum level—all we have are "overpowered cannons." We can use all sorts of tricks to minimize the effects of this, but we cannot make our observational "cannon" fire anything else except "cannon-balls."

The Heisenberg Uncertainty Principle

The net result of this inability for the observer to not interfere with the outcome of quantum events leads us to the *Heisenberg Uncertainty Principle*. In essence what this principle states is that since light arrives in "packets" or "quanta" of energy via photons, making a measurement will *necessarily* alter the object which is being measured. Specifically, the Heisenberg uncertainty principle refers to the measurements of the momentum (mass times velocity) of particles and the positions of particles. Because of the effects of using our "cannon-balls" for measurements, there is always a certain degree of uncertainty in momentum and position that cannot be avoided. It is important to realize that the "kicks" that observations give events are not the only grounds for the uncertainty principle; it is an intrinsic limit on the amount of information we can gather at the quantum level—even in principle—that we can know about a quantum object, regardless of if an observer is present or not. Another description of the uncertainty principle is found in The New Quantum Universe (Tony Hey & Patrick Walters—pages 22-23) which is profitable to share: "...the uncertainty of the *position* measurement...and the uncertainty of the *momentum*...are inextricably linked together....To determine the *position* very accurately it is necessary to use light with a very short wavelength since the wavelength of the light determines the minimum distance within which we can locate the particle...But, such high frequency [and wavelength] light will arrive in photons with a very large energy and give the quantum system a very large kick...If we want to know the *momentum* very accurately, we must give the system a very small kick...this means using light of a low frequency [you need to use low-frequency, long wavelength light to give the particle more *time* to move through space to more accurately determine it's mass and velocity, since momentum is mass times velocity]. Low frequency means long wavelength and this in turn means a large uncertainty in the measurement of *position!*"

This uncertainty is present on all scales; there are no stipulations that are bound to the quantum scale. However, in the *macro*scopic world, such effects are negligible and we are able to quite safely get away with our everyday "approximations" of events despite our "meddling" with measurements and observations. For larger objects, such as a car, where we want to know its exact position, the effects of the uncertainty principle amount to less than a *billionth* of an inch. Nevertheless that billionth of an inch is information that is permanently kept secret by nature's hand. Mathematically, the uncertainty principle is expressed as:

$$\Delta x \Delta p \geq \frac{\hbar}{2}$$

Where Δx is the *standard deviation* of the position of a quantum object (or in other words—how much the position data is "spread out") and Δp is the standard deviation of the momentum of an object, and $\hbar$ (called h-bar) is Planck's constant divided by 2π (the value for $\hbar$ is 1.05×10^{-34} Js [joule-seconds]). This equation expresses that the uncertainty in position times the uncertainty in momentum *must* be greater to or equal to h-bar divided by two. We can express this uncertainty in terms of energy and time as well:

$$\Delta E \Delta t \geq \frac{\hbar}{2}$$

Where ΔE is the uncertainty of the energy, and Δt is the uncertainty in the time. As we shall see shortly, this representation of the uncertainty principle provides an explanation of *virtual particles* in particle physics.

Uncertainty Principle Equates to Probabilities

The uncertainty principle has far reaching implications to our understanding of quantum events and quantum objects. We have established that there is a limit to the degree of certainty of momentum or position that we are allowed to achieve by our observational perturbations. I do not want this limitation to give a jaded view of classical mechanics in light of quantum mechanics. Classical mechanics is the foundation upon which we were allowed to get to the point of even being able to conceive quantum theory. Beyond this, classical mechanics still calculates macroscopic events to a remarkable precision that is well suited for our everyday needs. All those tiny approximations that only show up on quantum scales get averaged out in the trillions of atoms that make up every day objects that are interacting. The majority of our technologies are based on the principles of classical physics. Only when velocities get extreme, or scales shrink down to the quantum levels, do we need to invoke the new physics of relativity and quantum physics.

Despite the extraordinary precision to which quantum has been verified, there is still a hard fact that is difficult for people to accept: Regardless of how fine a detail we are able to observe, we are in fact not really observing the true nature of a system, we are seeing an approximation of that system. We can dial in our accuracy and keep adding those decimal places to the right, to get that extra 0.0000001, but at some point we will always run out of accuracy as dictated by the uncertainty principle. What we are really dealing with in our measurements is the *probability* that an event will occur. In reality there is an infinite amount of possible outcomes for events. Quantum physics states that the probability of an event that you are trying to predict must take into account *all* the possible outcomes for that event.

Virtual Particles and Renormalization

The intrinsic fuzziness that is built into quantum physics from the uncertainty principle has a strange implication when studying subatomic interactions. More specifically, the uncertainty principle shows a necessary amount of unpredictability in all interactions is required—period. During the smallest slices of time, Planck time, a particle has a probability of being able to do things that would normally not be allowed by the laws of nature. For instance, an electron can momentarily emit a photon and then immediately re-absorb it before nature "knows" a conservation law has been violated. Such a photon is known as a *virtual particle*, and this type of interaction is not limited to merely photons, it is thought that any particle can be "virtually" created this way. While this seems like a strange concept, and indeed it is, in the current lines of quantum physics it is believed to actually happen (in a nutshell—the mathematics behind quantum physics predict it, and to date, no experiment has *ever* contradicted a prediction of quantum physics). Electrons for example are thought to be surrounded by a cloud of these virtual photons. At greater distances, this cloud of virtual photons is not significant mathematically and can be disregarded; however on probing closer and closer to the "center" of an electron, these virtual photons interactions become more important and must be included in the quantum mathematics. The problem is that when you do this, the mathematics start to break down and spit out infinities for solutions, which obviously cannot be correct. Physicists came up with a mathematical trick called *renormalization* to solve this problem. In essence, renormalization divides each side of an equation by infinity, in order to cancel them out mathematically, leading to a finite answer. In Paul Halpern's book <u>The Great Beyond</u> (pages 208-209), he provides a clever way of thinking of renormalization in terms of accounting: "Imagine a business, Quantum Enterprises, that starts out with $100,000 in its till. Every day the company gains $1,000 but also has to pay $1,000 in expenses. This exact balance of earnings and losses continues indefinitely. With concerns about its long-term future, the company calls in two different accounts (for independent estimates) to calculate how much money it will have many years down the road. The first accountant is not too bright. As a first step he decides to compute the gains. He begins to add up all the earnings for not just year ahead but centuries as well. As he keeps entering figures into his calculator, the total gets greater and greater. Eventually, after he has computed the total for many, many centuries, the calculator overflows. 'Infinity.' he writes down as his answer. Then he decides to subtract each day's losses from the total. 'Infinity minus $1,000 is still infinity,' he writes down again and again. When it comes time to make his report, he announces, 'I'm pleased to tell you that your company has the potential for infinite profits, if you continue your current policies indefinitely. The second accountant is much more clever. She groups the figures in such a way that indicates their balance. Clearly the $1,000 lost subtracted from the $1,000 gained each day yields $0 per Diem. The initial $100,000, plus an endless series of zeroes, just makes $100,000. Presenting the company with this finite figure, she offers them a more realistic view of their long-term prospects. Such is the superiority of 'renormalization'". Ideally, physicists would prefer to not use such mathematical "trickery" to come out with finite answers (after all, in math class we are all taught never to divide by infinity as it is undefined mathematically). However, by using renormalization in quantum electrodynamics and other theories to precisely cancel out the values that are heading towards infinity, this technique agrees extremely well to experiment. If it was not for this tight agreement to experiment, renormalization would have to be disregarded.

If the idea of virtual particles still seems confusing, let us now explore our energy and time representation of the uncertainty principle:

$$\Delta E \Delta t \geq \frac{\hbar}{2}$$

If we look at this equation we can see that as the uncertainty in time t gets smaller and smaller, the uncertainty in energy E must become greater and greater to keep the product of the time and energy certainty equal to or below h-bar divided by two. So, if the smallest quantum of time, known as *Planck time,* is inserted in for t, then the uncertainty in E can be extremely large. In fact, for such short intervals of time, the system can <u>momentarily</u> have enough energy to create a whole particle into existence by means of the equivalence of mass and energy ($E=mc^2$) without violating the uncertainty principle or the law of conservation of energy. Essentially, this equation tells us that if things happen incredibly fast—particles can pop into existence and then pop out of existence before Mother Nature can do anything about it.

Quantum Probabilities Waves

All quantum objects can be thought of as both particles and waves. If you are trying to measure wave-like features then the quantum object being studied will exhibit wavelike behaviors. Conversely, if you are trying to measure particle-like qualities of quantum objects, they will accommodate as well. This statement is hard to grasp, and goes against the grain of our everyday experience (as much of quantum theory does).

To help explain this, let us invoke another experiment. Let us imagine a laser pointing at a solid sheet of metal that has two closely spaced parallel slits. These slits are so close together that they are approximately the same distance apart as the wavelength of our laser light. Behind our metal plate we have a screen made of phosphorus, which will chemically react with every photon that hits its surface, making it glow. We have a camera positioned nearby as well, in case we get the inkling to capture the results.

If we turn our laser on, it will start to fire coherent photons (that are all "waving" in sync—also known as monochromatic light, with the "mono" or "one" stemming from a laser emitting photons of *identical wavelength* and hence identical color [if in the visible part of the spectrum—which is the only region where the term "color" has a literal meaning]) towards our metal plate, upon reaching the plate, the majority of the photons will be absorbed or reflected back by the plate, however a few will make through the two narrow slits. What happens to the photons that make it through our device and impact the phosphor screen? They will produce an *interference pattern* consisting of bright and dark bands. As with water waves and sound waves, electromagnetic waves can amplify or cancel each other out depending on what part of the wave cycle they are on when they hit our phosphor screen. This happens because it takes slightly different amounts of *time* to cover the distance from one slit, as opposed to the other slit (just as crossing a street diagonally increases the distance, and takes more time than going perpendicular, or straight across the same street). These slight differences in time for the photons coming through our slits are enough for them to be at different parts of their wave cycles when they impact the surface of the phosphor screen. If a photon from slit #1 is at its *peak*—or highest energy level—and interacts with a photon from slit #2 that also happens to be in its peak as well, they will combine amplitudes (energy) and make a much greater combined "impact" on the phosphor screen (*constructive* interference) and produce a *bright band* on the screen. Where the photons are 180 degrees out of sync from the different slits, the photons from both slits cancel each other out (*destructive* interference) and impact the phosphor screen with little or no energy left to impact and initiate the chemical reaction with the phosphor screen. This scenario leaves a *dark band*. This type of behavior is characteristic of waves.

Now let us consider another setup for our experiment. Instead of letting our laser emit its typical trillions of photons per second, let us attach a device that restricts the number of photons coming out in an arbitrary amount of time to be only *one*. Our camera will come in handy now, as we can secure it in a position facing our phosphor screen, turn off the lights, and leave the shutter open to record how our individual photons are "piling" up on our screen over time. With our new experimental setup in hand we turn on our laser that is modified to emit one photon at a time, and let it run while we step out for coffee and a bite to eat. Upon our return, we turn on the lights, and turn off our laser, and develop our film. What will we see? The surprising answer is that we will see the exact same pattern of dark and bright bands (interference pattern) as we did before. How is this possible? The answer is that instead of thinking of photons (or any quantum objects) as "waving" in a traditional sense, such as water waves, or sound waves; where water or air are transferring the energy of the wave through that propagation medium to another location. Quantum objects such as photons travel on *probability waves* that are *independent* of a medium. In other words, *photons do not need a medium to propagate through, they are self propagating*. If we remember back to the uncertainty principle, this makes more sense. We can never know the exact position of a quantum object in space and time; we can only know the *probability*. The probability wave equation is the same as any other wave equation graphed out, with peaks and troughs extending through time. However for a probability wave function the peaks represent the greatest probability of finding that quantum object at a certain point in space at a certain point in time. Likewise, the trough of the wave function represents where the probability of finding a quantum object in a certain point in space and time are the lowest or zero. We can never predict where any one photon is going to "land" on our screen. We can only calculate the probability of where it *could* land based on the probability wave. Below is a sample example of a probability curve to graphically see what I am describing, along with a figure outlining the double-slit experiment as performed with water waves, and electrons.

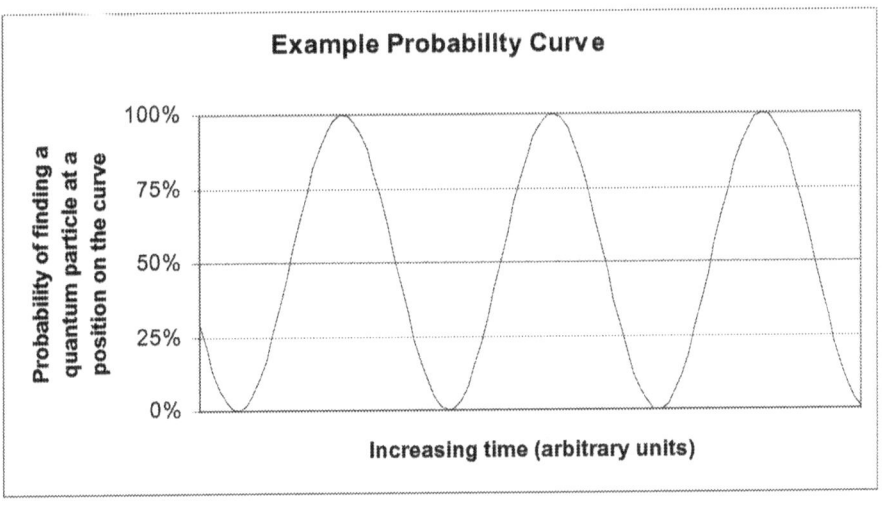

The double slit experiment where a water wave source S emits waves that impact wall A which has two narrow slits in it. After passing through the slits at A, the waves split and interfere with each other on their was to the detector that records the impacts of the wave at point B. Where two wave peaks (as indicate by the arcs) intersect, there will be a greater impact at point B—making the black bands. This pattern is what will make up the interference pattern that is indicative of wave behavior.

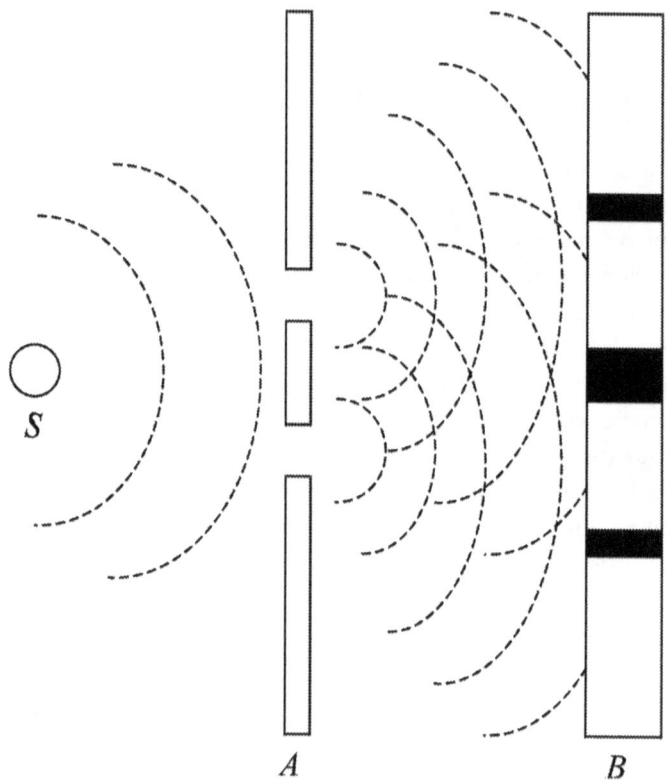

The same double slit experiment as performed with electrons being emitted from an electron gun at point S instead of water waves. Despite electrons being individual particles, we observe the same results as we did with our water waves. Even if you slow down the rate the electrons are firing to a point where they are being emitted individually, the same interference pattern at point B will still build up over time. This behavior is a mystery that lies at the heart of quantum physics.

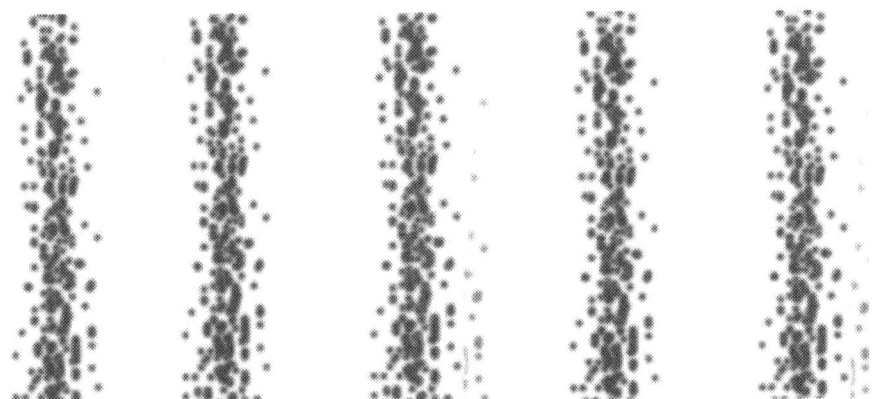

Sample close-up of the electron double slit experiment results. If you magnify the interference pattern you can still see the impacts of the individual electrons on the detector at point B. Over time, the individual electrons built up the pattern above. It is as though the electrons leave point S as points, interact as waves while on the way to point B, where they once again impact as points once again, but in the pattern of wave interference.

Schrödinger's Wave Equation

Erwin Schrödinger formally introduced wave equations that forever changed the world of physics. His equation, the *Schrödinger wave equation* (the one-dimensional, time-independent version) is mathematically given by:

$$\frac{d^2\varphi}{dx^2} = -\frac{2m}{\hbar^2}[E - U(x)]\varphi(x)$$

Where $\phi(x)$ is the wave function, m is the mass, $\hbar$ is Planck's constant divided by 2π (h-bar), E is the total energy (of the particle), and $U(x)$ is the potential energy function (of the particle). For the scope of this book, we will not explore the mathematics behind this function; however it is nonetheless an important enough equation to be familiar with, even without the specifics of the mathematics. Schrödinger's equation throws out the old concept of atoms as being tiny little solar systems as envisaged by the Bohr model, with electrons circling the nucleus in "orbits". Schrödinger's atoms were now thought of (correctly) as probability shells, or in the case of electrons—probability orbitals. Neil Bohr's trajectory-specific model of the atom was replaced by probabilistic wave functions, where the orbits were "soft" shells of 80 to 90 percent probability of finding an electron at a particular point. Schrödinger's probability wave equation set the stage for much of quantum physics. It is worth noting that the mathematician Oskar Klein developed an earlier version of the Schrödinger equation, however he was too ill at the time he discovered it to see it published (Klein is best known for this efforts, along with Theodor Kaluza on higher dimensional unification theories that become a cornerstone for modern superstring theories and M-theory, which will be explored later in this book). Additionally, the original construct of Schrödinger's equation was a non-probabilistic one (deterministic). It was actually Max Born who introduced the notion of the wave actually being a probability wave. Upon hearing about this Schrödinger was furious at the thought, and did not want to entertain the non-deterministic view of electron orbitals.

The Duality of Light and Matter Waves

An often confusing aspect of light is that is exhibits characteristics of waves and particles. Put another way, light travels like a wave and exchanges energy on small scales with other quantum objects like a particle. This duality of light is a well understood part of quantum physics. A more hidden reality that is often not explored is that this wave-particle duality is not limited to light. Louis-Victor DeBroglie discovered that *all* objects, *regardless of size*, exhibit a wave-particle duality in the form of *matter waves*. De Broglie's equation to describe the wavelength λ of any matter wave is:

$$\lambda = \frac{h}{p} = \frac{h}{mv}$$

Where *h* is Planck's constant (6.6260755 x 10^{-34} J s), *p* is momentum, *m* is mass, and *v* is velocity. We have verification of this based on our previous example of our photons producing interference patterns that are characteristic of waves. Based on DeBroglie's equations, we find that for *macroscopic* objects, the wavelength is so small as to be undetectable; so once again we can get away with approximating objects as not having a wave-particle duality on macroscopic scales when in fact they do. The following table summarizes the De Broglie wavelength for object's with an electron's mass, all the way up to an object with the mass of 100,000 kilograms:

	mass (kg)	velocity (m/sec)	De Broglie Wavelength (m)
*electron	9.10908E-31	299,792.458	2.43E-09
	:	:	:
	0.00001	1,000	6.63E-32
	0.0001	1,000	6.63E-33
	0.001	1,000	6.63E-34
	0.01	1,000	6.63E-35
	0.100	1,000	6.63E-36
	1	1,000	6.63E-37
	10	1,000	6.63E-38
	100	1,000	6.63E-39
	1,000	1,000	6.63E-40
	10,000	1,000	6.63E-41
	100,000	1,000	6.63E-42

Quantum Tunneling

Taking into account all the possible probabilities represented by a quantum event's wave function leads us into some mysterious waters, for there are miniscule probabilities for just about any sort of crazy thing you are trying to measure to actually happen. For example, in L.E. Lewis Jr's book, Our Superstring Universe (page 68) he briefly explains how there is a measured probability according to quantum mechanics of a person being able to pass through a solid object once every 15 billion years, if making attempts every second. The reason why this probability is so extraordinarily tiny is because for our trillions and trillions of atoms to pass unimpeded through a solid object, such as a wall, all of those atoms would have to line up just right so as to not interact with any of the atoms of the wall. Another example is that there is an extremely small probability that all the quantum objects that make up this book you are reading will all move in the same direction at the same time, thus making the book "jump" right out of your hands. The reason that there is even a possibility of passing through a wall is due to the fact that matter is mostly made up of *empty space*. Quantum objects such as the protons and neutrons of atomic nuclei, or an atoms associated electrons, are a tiny part of the overall volume that the probability shell of the atom as a whole encompasses. The vast majority of the volume of what we call an "atom" is empty space. To get a better sense just how much empty space there really is in an atom, let us look at the scales involved. An atom has a diameter of about 10^{-8} cm. The nucleus of the atom, which houses practically all of the mass of an atom, is only about 10^{-13} cm. If you scaled up the whole diameter of an atom to the size of a large room, its nucleus would be a tiny point that could be barely discerned with the naked eye (Richard Feynman, Six Easy Pieces, page 34).

So, for an object the size of a human, we are looking at a probability of once every 15 billion years if we wish to walk through a wall. What if we wanted to see an electron or photon travel through a solid object, how long would we have to wait? As it turns out—not too long. Since photons and electrons are easily produced in such massive numbers, even if a given event has a small probability of occurring, we can squeeze millions of "attempts" into a small interval of time, thereby greatly reducing the amount of time to wait for that improbable event to occur.

An experimental example of this in quantum electrodynamics is explained in Feynman's QED (pages 20-21). When you take a beam of *monochromatic* laser light (coherent light of one wavelength) and bounce it off different thicknesses of glass and detect how many photons are reflected off the glass. As it turns out, something very counter-intuitive occurs. As you increase the thickness of glass, the number of photons from the laser that reflect off the glass and reach the detector in a given length of time *cycles* between a high and a low percentage *indefinitely* (according to Feynman this cycle of reflectivity has been experimentally verified for 100,000 cycles—which corresponds to glass that is 50 *meters* thick). Let us say that we start with an arbitrary thickness of glass and we register 8 out of every 100 photons that the laser emits as being reflected by the glass and reaching the detector, or 8%. As we gradually increase the thickness of the glass, we notice that our percentage of photons being detects goes *up* as well (despite the increased thickness of the glass), until it reaches 16%. If we increase the glass thickness more, our percentages start to go *down* again—15%, 14%, 13%....until we reach a point where it is at zero percent—where no photons from our laser are getting through. If we keep thickening our glass from this point, our percentages start *rising* again. This rising will continue for our thickening glass until we reach 16% again, where it again starts to fall. If we repeated the above experiment with a more energetic laser that emitted shorter wavelength photons, we would observe the same cycle of zero to 16% reflection rates, however these reflection rates would be more rapid; that is, they would be more sensitive to the thickness of the glass (still cycling between zero and 16%). This strange effect is called *quantum tunneling*. The reason why we do not see this strange effect in our everyday light is because normal lighting is not monochromatic and is made up of photons of many different energy levels. Each of these energy levels interacts with a piece of glass in a different way, falling somewhere on our zero to 16% reflection spectrum. However, *overall* the *average* of all of these different wavelength photons interacting with the glass comes out to a consistent transmission number—8%. Quantum physics dictates that we cannot predict where a single photon (or any quantum object) will "land" once it leaves the laser. We can only calculate the *probability* of it landing in a certain spot.

Two more good examples of how to look at quantum tunneling are presented in John Gribbin's book The Search for Superstrings, Symmetry, and the Theory of Everything (pages 37-38) in explaining how an alpha particle can escape from an atomic nuclei in certain radioactive material (an alpha particle is a helium nucleus—consisting of two protons, and two neutrons). "The alpha particle sits inside the nucleus, and we can imagine it as being just inside the rim of a volcano. If the particle were just *outside* the rim, it would 'roll away', and be ejected by the force of electric repulsion [what we would see as radioactive alpha particle emission]…An alpha particle associated with the nucleus has a very well defined momentum, as does the nucleus itself. But that means its position must be uncertain [as stated by the uncertainty principle]. Even though an individual alpha particle does not have enough energy to climb over the inner rim of the volcano and escape, it is not *inside* the nucleus, in the everyday meaning of 'inside'. Uncertainty implies that there is a finite, and precisely calculable chance that the particle is actually outside the nucleus…Some particles do find themselves outside the nucleus, take note of the fact, and rush away, just as if they have 'tunneled' through the intervening barrier. It is exactly as if you took some dice and rattled them in a cup until suddenly one of them appear outside the cup, rolling across the table. And if Planck's constant were big enough, that is how dice would indeed behave in the everyday world…Or think of it [tunneling] in terms of energy. The particle needs more energy to climb over the 'rim' of the potential barrier…For a brief enough instant of time [Planck time], it might, for all the laws of physics know or care, have that extra energy. And if it does, once again it is off and running. It doesn't matter that it has to 'give back' the energy it borrowed from uncertainty [from the law of conversation of energy]…because by then it has escaped down the hill on the other side of the barrier."

How are these types of events possible? This question has a simple answer—*we don't know*. We do not know how a photon "decides" whether to go through the glass (or other barrier) or get reflected. Quantum physics dictates that in the double-slit experiment we cannot define an electron or photon as a wave or particle until we *observe* the system. Until an observation is made, then we must assume that the photon or electron is in a *superposition* of all possible states of both as defined by the Schrödinger wave function. As hard as it is to take such a strange behavior as not having an underlying meaning, that is how it is. Quantum physics allows us a tool to predict with great precision a way to calculate and quantify these results in the form of probabilities; however it does not provide us with the reason why.

Quantum Objects as Probability Shells

I have periodically been referring to atoms as *probability shells*. Now we are in a position to explain this statement with the topics we have covered. Let us first consider a mental example of an "event" for reference. Imagine you have a big salad bowl and the b-b we used to try and slow down that bowling ball earlier. If you hold that b-b against the inside of the bowl and let go, what happens? Obviously, the b-b will roll all around; tracing out arcs as it speeds towards the center of the bowl, overshoots, and then rolls up the other side to a slightly lesser height, and then roll back towards the center of the bowl, overshoots…and so on. Over time, the b-b will eventually stop moving and come to rest in the center of the bowl from frictional forces between the surface of the b-b and the inside of the bowl, and with the air.

Now let us explore a quantum-sized example with our knowledge of the uncertainty principle. Instead of using a b-b, we are going to use an electron. We can retain the use of our bowl from our previous example for this mental experiment. If we drop our electron into the bowl, what happens then? As with our b-b, the electron will trace out arcs and ellipses as it accelerates towards the center, overshoots, and then rides up the wall of the other side of the bowl…etc as before (ignoring electromagnetic forces). However, this electron will never completely stop moving in this example. It would settle down into an area that is very close to the center of the bowl as the b-b did, but it will always be jiggling around and never completely come to rest. Even if you left this electron in your bowl for weeks, months, years…it would continue to have these random jiggles at the center of the bowl. Why is this? The answer lies in our uncertainty principle. We are unable to know *exactly* the momentum and position of a quantum object like an electron *simultaneously*. In order to measure with high accuracy the *position* of our electron in the bowl, we necessarily will give up some accuracy on its *momentum*. Likewise, if we try to accurately measure the *momentum* of our electron (mass multiplied by velocity) we need to give up information on its *position*. The electron cannot stop jiggling around because it cannot be at a state of complete rest at the bottom of our bowl since that would violate the uncertainty principle as it would have an exact location and momentum.

This constant jiggling is based on sound mathematics and is the principle behind the concept of quantum objects being probability shells. Just as our electron in our bowl, all quantum objects can never be truly at rest (this is in part why the lowest possible temperature—absolute zero, is never totally achievable experimentally). We can never know *exactly* where quantum objects are located; we can only calculate a "cloud" or "shell" of where they *probably* are. In quantum physics, this lowest energy vibration that cannot be removed is called *zero point motion* or *zero point energy*. This is the lowest possible energy state a quantum system can achieve without violating the uncertainty principle.

Many textbooks still refer to an atomic model that depicts a small, dense, nucleus surrounded by a cloud of orbiting electrons. Based on what we have explored previously, this is not the case. Quantum objects cannot be actualized more than their presence in a small shell of that indicates their probable location. However, despite some textbooks hanging onto the older atomic model (perhaps as it is more intuitive and easier to understand than the true nature of atoms) it still raises an interesting question: What keeps those orbiting electrons from falling into the nucleus? To answer this, we once again must turn to the uncertainty principle. If an electron fell directly onto an atomic nucleus, then we would know its position to a high order of precision, which, according to the uncertainty principle, would require that our electron have a very large uncertainty in momentum. This momentum is equivalent to the electron having a large amount of kinetic energy (energy of motion). This increase in momentum from having a more precise location at the nucleus will be enough momentum for it to break free of the nucleus. This extra momentum will always ensure that electrons do not fall into the atomic nucleus, so they are confined to stay in their cloud, each wiggling in their own smaller probability shells in accordance to their obedience to the uncertainty principle.

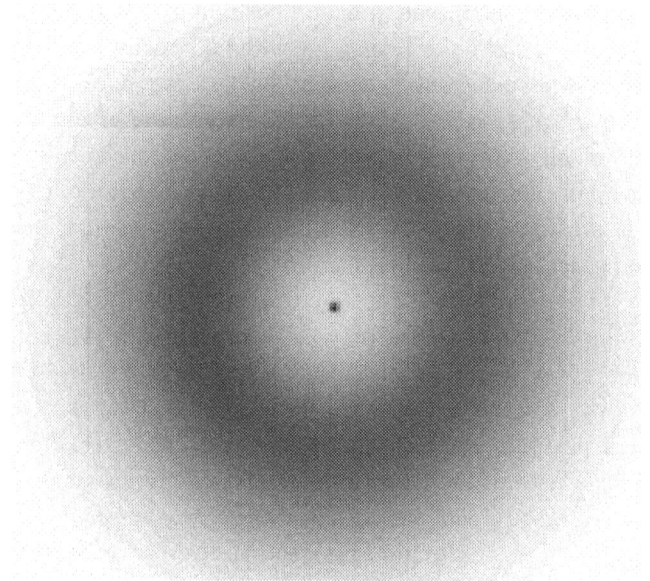

Modern concept of the atom based on quantum physics. The concentrated mass of the nucleus is represented by the tiny dot in the center. The probability cloud of where the electrons can be found is represented by the grey cloud ring surrounding the nucleus.

Photon Travel at the Quantum Level

Quantum physics shows us that the probability of any event also take into account all the possible outcomes of that event in order to accurately predict that probability. So what does this mean for photons traveling at the quantum level? For an example, let us imagine a ray of light traveling from a distant star to our eye. If you could trace the path of that ray of light back to the star, it would be a straight line, correct? For everyday scales, we can safely answer "yes" to this question (we are momentarily ignoring the effects of warped space-time from relativity in this example). On quantum scales, quantum theory has shown us that light does *not* travel in straight lines. Instead, it travels in a small "core" of space that consists of the probabilities of *all* the possible routes that stars light could have taken to reach your eye, whose combined probability focuses into a small area, or core. Richard Feynman has a good description of why light approximates a straight line when traveling between two objects in his book, <u>QED (page 54)</u>: "For each crooked path [between the two objects that photons of light are trying to travel] there's a nearby path that's a little bit straighter and distinctly shorter—that is, it takes much less time. But where the paths become nearly straight…a nearby straighter path has nearly the same time…that's where the light goes. It is important to note that the single…straight-line path…is not enough to account for the probability. The nearby, nearly straight paths…also make important contributions. So light doesn't *really* travel only in a straight line; it "smells" the neighboring paths around it, and uses a small core of nearby space. (In the same way, a mirror has to have enough size to reflect normally: if the mirror is too small for the core of neighboring paths, the light scatters in many directions, no matter where you put the mirror.)"

Superluminal Velocities—was Einstein wrong?

Einstein's theory of general relativity was perhaps the most difficult theory ever envisioned by a single human mind. A core concept of relativity is that the speed of light is the speed limit of objects traveling *through* space *in a vacuum* (light does slow down in different density mediums and gives us our familiar effects of iridescence on soap films, rainbows, indexes of refraction…etc). However, Einstein found that nothing can travel *faster* than light—186,000 miles per second is the maximum speed allowed…Right?

This actually is not true according to quantum theory. Just as classical Newtonian physics on larger scales work well despite the approximations they make in light of the scrutiny that relativity brought to bear on them, so to has this happened with relativity itself in regards to quantum physics. As it turns out, quantum theory predicts that the speed of quantum particles can *exceed* the speed of light over the tiniest of distances (Planck distance). The reason for this stems from our previous discussion of how quantum physics is based on *probabilities*, and that the probability of an event is based on the total of *all* the possible outcomes of the event. As it turns out, on quantum scales there is a probability that must be taken into account for photons that momentarily travel *faster* or *slower* than the speed of light. These variations in photons velocity probabilities happen continuously at the quantum level, where the distances involved are small enough for these to become significant. As you zoom out, and start examining on larger scales and distances, these probabilities of velocities faster and slower than the speed of light get averaged (or cancelled) out to what appears to be a the constant velocity of light. So Einstein is still correct: The *average* speed limit of light over *longer distances* is still 186,000 miles per second. On all but the tiniest of quantum scales we can achieve the accuracy we need by approximating light speed as a fixed value. Only in the realm of quantum physics do those small probabilities for superluminal and subluminal velocities start to become important considerations. Just as Einstein's relativity transcends certain notions of Newtonian physics, quantum physics transcends certain notions of relativity.

Profound Implications—Is Reality Real?

The topics in the previous section are difficult to accept for some people, for it has a profound implication: If we can only approximate quantum objects as shells of probability in locations of space, then for all we know quantum object may not even exist. As strange as that sounds, with our current knowledge of science, we must allow it as a valid possibility. So if we cannot confidently state that quantum objects are what we call "real" objects in our everyday experience, such as book, or a cup of coffee, then what is the big deal? The big deal is that *everything* in the cosmos that we can perceive with our senses is built out of these same quantum objects. A quantum object, such as an atom of hydrogen, does not have different properties when being studied in a lab or as part of a human being's body—it is still the same hydrogen atom. *All quantum objects have the same properties of probability shells regardless of what larger structure they are incorporated into.* This means that we must allow the possibility that our whole cosmos as we know it, does not exist in the "real" sense, and our perception of "reality" is a falsity. This perception of reality is an old topic in philosophy, and from the scientific end it can be seen from other angles as well. Einstein for example, showed that the speed of light is the fastest speed through space that is obtainable. We can apply this speed limit to our everyday experience and note a similar falsity of reality as we do in quantum physics. For, if the speed of light is the fastest possible, and that speed is finite—which we know by experiment and have measure accurately, then it takes *time* for all the information that you *see* to become observable. In other words, whatever you are looking at is not in fact reality. You are in fact seeing what objects looked like *back in time* since it takes time for photons from those objects to travel over a particular distance to your eyes. And, beyond that, it takes time for the chemical reactions in your eye to register the arrival of those photons, and to send that signal to your brain to process.

Such concepts start overlapping ground that is covered in philosophy, and ultimately may never be able to reach closure with scientific faculties. Personally, I find it marvelous that quantum theory, the most extensively tested and successful theory created by humans, entertains and even invites such "absurd" trains of thought. It shows that even on the finest of scales, Nature remains veiled in mystery, and continues to spark our imaginations, and drive our curiosity. I do not think that the concepts of quantum theory should be viewed with surprise, for while it has many counter intuitive concepts nestled throughout it, it is very much in the lines of our everyday experiences. The concepts of uncertainty are dealt with daily. Our entire lives are unpredictable from one moment to the next. It should not be a surprising discovery that the most successful theory in physics is ultimately based on Nature being unpredictable as well. Conversely we should allow ourselves a moment of amazement that on quantum scales—the building blocks of everything we are and will ever be—Nature has bestowed her intrinsic qualities in each of us.

Quantum Controversy

While quantum theory is indeed considered one of the two pillars of modern physics, along with relativity, it is still a theory. In reality, it is important to realize that no *matter how successful a scientific theory is, on fine enough scales, it will always be an approximation of the truth.* There is never going to be any way to break open the "watch" of nature and see how everything works. Einstein scoffed at the ideas underlying quantum physics, despite him discovering the quantum of light, the photon, and making important contributions towards quantum theory's development.

Einstein is not alone in this mindset. Despite decades of successful predictions, and experimental verification of quantum theory, there are still physicists who believe that quantum mechanics intrinsic "fuzziness" is a consequence of a shortcoming of our understanding of nature. There are still physicists (albeit a minority) that believe that the universe is ultimately governed by classical laws that we as of yet have not discovered. In this view, quantum physics would merely be a façade to some deeper classical laws that are currently unknown, or whose information has been washed out with interactions with other systems. For example, in the September 2004 <u>Scientific American</u> (pages 88-91) there is a good analogy presented to show this type of information loss due to dissipative forces such as friction. The analogy is to imagine a baseball pitcher throwing a series of baseballs off of the top of a tall skyscraper. At the bottom of the sky-scraper, we have an array of instruments to accurately measure each baseballs speed just before they impact on the ground. The pitcher throws the first ball straight out off the top of the building at 40 miles per hour. This ball will go rel-atively straight for a certain distance, and then gravity will start curving it back down towards the ground, accelerating it as it goes. Eventually, the baseball will reach a maximum velocity (terminal velocity) due to wind resistance on the way down. Before impact, we measure its velocity and determine it to be 80 miles per hour. Now, let us have our pitcher throw a fast ball, whizzing off the top of the building at 90 miles per hour. Again, the balls trajectory will curving towards the ground and it will accelerate under gravity until reaching a terminal velocity that was identical to the 40 mile per hour baseball. On the ground, we would still measure it as traveling 80 miles per hour just before impact. The loss of informa-tion on the initial conditions due to wind resistance is obvious. In fact, if the building that our pitcher is throwing from is tall enough, it does not matter how fast or slow, or at what angle they throw baseball from (a ball thrown much faster than 80 miles per hour, will slow down to 80 miles per hour due to frictional drag due to the same wind resistance). On the ground, we will always indicate the same reading of velocity prior to impact. All of the information of the initial con-ditions of the system is lost as energy is dissipated via friction into the air sur-rounding our baseballs.

In this example, all of the balls eventually equalize to the same terminal veloc-ity of 80 miles per hour due to the balance between air resistance and the acceler-ation due to gravity. In physics (and chaos theory) this terminal velocity is said to be an *attractor*, for as we have seen, regardless of the initial conditions we give the baseball to hurl it into freefall, the velocities all eventually converge, or "attract" to a single value. Another possibility is that of higher dimensional considerations. More specifically, there is evidence that classical physics-based systems operating in higher dimensional spaces can assume many quantum physics like properties when observed in four space-time dimensions.

Clearly, this is still a very much open debate, and to reiterate, the physicists that hold these views are in the minority. However, we must hold a door open for these types of viewpoints, as they could still ultimately be truths. Many aspects of physics continue to reveal more layers to Nature's onion. Often in the past, what were once thought to be well understood concepts turn out to be still quite nascent upon new experimental evidence.

The Arrow of Time

Our above example of dissipative forces is a good tangent to a paradox that is often discussed in physics: that of the apparent *arrow of time*. In our every day lives, we take this paradox for granted. We know that we can remember events from the past; however we are unable to "remember" events in the future. If you see a broken glass on the floor, you know that it will not suddenly un-break, with all of its constituent pieces forming back into an unbroken glass. It may seem like an obvious concept to point out; however in the mathematics that governs our laws of physics, we do not see this direction or *arrow* of how events are order. At the fundamental level, the laws of physics as we know them are *time symmetric* (or time-reversal invariant). That is to say, the laws of nature as uncovered by physics do not treat the past any differently than the future. Obviously, this is not true in our everyday world, and this is the paradox. Everything around us indicates that time has a definite arrow that points from the past to the future; where as the physical laws that are underlying those same everyday experiences seem to be time symmetric and treat the past with the same integrity as the future.

This discrepancy is due to the sheer number of atoms and interactions involved in our everyday scales. Even the smallest of objects that we can see with the naked eye consists of trillion of individual atoms, each of which are obeying the physical laws of nature. Once we cross the boundary from microscopic to macroscopic then dissipative forces such as friction and viscosity introduce the familiar arrow of time. On macroscopic scales, the second law of thermodynamics (sometimes referred to as the law of entropy) brings forth the arrow of time, making events happen in a certain sequence. *The second law of thermodynamics states that, all things being equal, a system will tend to evolve from a state of low entropy towards a state of high entropy.* Things evolve more from orderly states to disorderly states over time. Again, this makes sense to us. If we become fed up with cleaning a room of our house and stop, that room will not become cleaner and cleaner as time goes on. It inevitably will become dirtier. *Entropy* is the amount of disorder that is contained in a system. A pristine, unbroken egg is an example of a low entropy state since there is only one way to arrange an egg when it is unbroken. A broken egg is an example of a high entropy state, since there are many different ways to break an egg. Another way to envision entropy is to think of it how noticeable a rearrangement of a system will be. For instance, if you have a pile of flash cards with numbers on them, that are all in a sequential pile, any rearrangement of the cards will immediately be noticed (hence this is a low entropy configuration). Conversely, there is a multitude of ways that we can rearrange the cards if they started out of order that will go unnoticed (hence a highly entropic configuration). Despite what has been presented above, on atomic and subatomic scales, the fundamental laws of nature still seem to treat the past and the future equally.

8

Fractals—Nature's Geometry

Practically everyone is familiar with Euclidian geometry. This is the mathematical world of planes, circles, squares, rectangles, triangles, spheres, cubes, pyramids...etc. Euclidian geometry is invaluable for describing things that humankind creates, and many of the simpler structures of this type of geometry can be built upon each other to grasp a geometric understanding of intricate systems. Despite this, Euclidian geometry deals with 1-dimensional, 2-dimensional, and 3-dimensional objects *only* (e.g. a line, square, cube, respectively). For natural systems, such as streambeds, coastlines, trees, mountains, continents, galaxies, and clusters of galaxies, Euclidean concepts are not ideal. Natural systems are the result of evolutionary progress. You do not see triangular trees, cubic continents, rectangular clouds, or perfectly circular galaxies.

Benoit Mandelbrot, a mathematician who worked for IBM in their research lab, noticed the same limitations of Euclidian geometry in describing these natural systems as well. He developed a new type of geometry, called *fractal* geometry, which was designed specifically to describe the geometric features of natural systems. The core concept behind fractal geometry is that it utilizes *fractional dimensions*. Such a notion may sound absurd at first; however it actually makes logical sense. Technically, the concept of fractional dimensions has much more fidelity to how objects really are. This strange concept of a fractional dimension can be explained by us first asking a simple question: What is the dimension of a ball of thread? We will have Benoit himself explain this example via a quote from his book The Fractal Geometry of Nature (pages 17-18): "A ball of 10cm diameter made of a thick thread of 1mm diameter possesses several distinct effective dimensions. To an observer placed far away, the ball appears as a zero-dimensional figure: a point...As seen from a distance of 10cm resolution, the ball of thread is a three-dimensional figure. At 10mm, it is a mesh of one-dimensional threads. At 0.1mm, each thread becomes a column and the whole becomes a three-dimensional figure again. At 0.01mm, each column dissolves into fibers, and the ball again becomes one-dimensional, and so on, with the dimension crossing over repeatedly from one value to another. When the ball is represented by a finite number of atom-like pinpoints, it becomes zero-dimensional again."

The key point to remember from this is that the observed dimension of the ball of thread depends on the *scale* at which we are looking at it. We can apply this concept to *any* object, whether it is artificial or natural. Just as relativity states that different people will see the same event differently relative to their frame of reference; fractal geometry states that the observed dimension of any object is relative to the *scale* at which it is being observed. The fractal dimension of an object can be thought of as the amount of "roughness" a surface has at a particular scale. A straight Euclidian line has a dimension of one. A typical coastline, for instance, may have many jags in it that make its dimension fall somewhere between our line of dimension one and a plane of dimension 2 (e.g. 1.28, or 1.57 fractional dimension).

Let us expand on the idea of a coastline to bring to light another phenomenon associated with fractals: Near-infinite length in a finite space. Again, as with our concept of dimensionality changes with the scale applied to any object, this concept applies to any object as well. If you looked up the length of the border of the United States (the perimeter) in a geography book, it would have a finite value assigned to it. However this is not accurate based on fractal geometry. This inaccuracy makes sense if you think about it more completely. If you measured the perimeter of the United States from a satellite photograph, you would get a different finite answer than if you measured it based on a series of aerial photographs that are closer to the coastline and give more detail. Your aerial photograph analysis would work out to a larger perimeter than your satellite analysis due to the increased detail seen. If you hired a person of extreme patience to walk the entire perimeter of the United States with a pedometer you would get a greater value still, as that person will be able to account for more of the coastline on foot, walking in and out of more and more of the bays and coves on the ocean sides of the border. The point is clear: the finer the scales you use to study the United States border (or any border for that matter), the higher and higher the value for its perimeter will be. For all practical purposes there is *no limit* to the additions to length that get introduced at smaller and smaller scales, the perimeter hence becomes unbounded (heads towards infinity). To capture this more fully, imagine you have an ant walk the distance, and then a flea, and then a microscopic worm…etcetera. For each finer level of scale, the perimeter value would become correspondingly greater and greater. This principle can be applied to any object, since regardless of how smooth an object appears at a particular scale, there will always be a scale at which its individual components will start to become visible, thereby transforming the edge of the object from a Euclidian boundary into a fractal boundary that will increase its complexity and thereby the length of the objects perimeter.

Self—Similarity

Another concept that is central to many fractals is *self-similarity* on different scales. Let us consider our coastline example again from earlier. If you view a coastline from different scales, they will have similar features: bays, peninsulas, sub-bays…etc. You can observe the same features in the same approximate numbers on a range of scales—making it *self-similar*. In natural systems, this makes intuitive sense upon examination. A coastline is self-similar because the same processes of erosion are at work on every scale of the coastline. There is not a particular favored scale in which erosion is less effective than on other scales. Natural systems of all scales are ultimately evolving within the same cosmos. Self-similarity is an intrinsic quality in nature.

Nature's Feedback Loop

As mentioned before, Nature is evolution based. Ever since the big-bang, Nature has been evolving based on sets of rules. Essentially, everything in the cosmos is a *feedback loop*; where the present outcomes of any system (from the quantum to cosmic level) are driven by the results of the last output of the system. In a mathematical sense, you can think of the cosmos as one big equation that is evolving based on motions that we label as time. For every instant of time new numbers come pouring out of our master equation. The present state of our equation relies on using the results from the last instance in time as *input* for the current calculation of that *same* equation. This is the feedback or *reiterative* concept of all natural systems. In this line of thinking, it becomes once again clear that self-similarity in nature is inevitable. Analyses of natural systems on all scales are showing signs of self-similarity and association with fractal geometries (and after all this should not be surprising as fractal geometry was *designed* with natural system in mind).

Complexity Generators

A very interesting aspect of fractals is their ability to create infinite complexity out of very simple sets of rules in a finite space. The Mandelbrot set is a fractal set that was discovered by Benoit Mandelbrot. This set represents the most complicated set in mathematics; yet the rules that govern the behavior for this set are incredibly simple. Mathematically, this infinitely complex set is produced by re-iterating the following equation on the complex number plane:

$$z \rightarrow z^2 + c\,.$$

That is it. The rules for producing the Mandelbrot can be represented by a few lines of computer code that pick a number on the complex plane and place it into the above formula in such a way that the result, or output, of the equation gets plugged back into the equation for the next round of input. If after a certain number of cycles (or so-called iterations) of this, the number starts heading towards infinity, then that number is *not* in the set as it is unbounded. If however, after iterating this equation at a particular point on the complex plane, it remains bounded and does not race to infinity, then it is in the set, and gets assigned a color on the computer display to identify it. After picking enough points, a bug-like form emerges that is self-similar. However, despite this self-similarity, it does not repeat itself *exactly* anywhere. Similar appearing features of the Mandelbrot set, if scaled up to the same size on transparent paper, would not exactly line up. This set of numbers has been magnified via computer to the equivalent more than a billion times its original size, and the complexity never shows any signs of subsiding. This is a very powerful concept that such infinite complexity can be contained within a finite space. Often when studying natural systems, we run into increasing complexity on different scales, with little or no signs of letting up. For example, we see this type of behavior when examining matter. Matter is made of atoms, and those atoms are made of electrons, protons and neutrons, and those protons and neutrons are made of up three quarks each…etc. If superstring or M-theory turns out to be correct, then each quark would in turn be made of a tiny vibrating strand of energy.

More to Explore—Getting Hands On With Fractals

With the advent and advancement of computer technologies, fractals are becoming available to the general public. There are several freeware and shareware software packages that can be downloaded off the internet that provide a great way to visualize how complex and beautiful fractal objects can be. I have personally enjoyed using the freeware software ChaosPro, which, as of this books printing was available at http://www.chaospro.de. This program allows the user to explore many different types of fractals, in both 2 dimensions and 3 dimensions. It also has capabilities for creating 2D and 3D animation sequences. Below are sample images of the Mandelbrot set using this software.

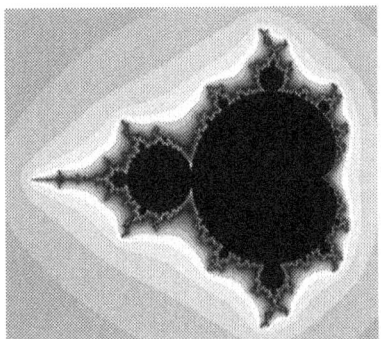 

The Large Scale Distribution of Galaxies—The Largest Fractal Set

As you remember from the first section of this book, we introduced the reader to the vast range of scales that are encountered in astronomy and cosmology. When we reached the scales beyond galaxies, we explored the large scale distributions of galaxies. This distribution is worth revisiting here as it is truly remarkable, and has some fractal type features that dovetail into our current topics. The key feature of the largest datasets of galaxy positions, when plotted out three dimensionally, is huge voids or bubbles that are almost completely free of galaxies. Along the edges of these bubbles, there are arcs and filaments of galaxies all grouped together. The exact reason for these types of distribution patterns in galaxies is not yet known. However, there has been research into the qualities of these voids and arcs, and it has been discovered that they are self-similar across a large range of scales, and seem to be fractal in nature. As mentioned in the first section of this book, the interested reader will profit by downloading the freeware software Galaxy Explorer at http://cas.sdss.org/dr3/en/help/download/ for a first hand look at what these structures look like when plotted in three dimensional space.

An interesting recent development about these galaxy distributions was outlined in <u>Sky and Telescope, May 2005</u>. Advanced statistical analysis of large-scale galaxies distributions reflects a "slight" tendency for galaxies to be 500 million light-years apart from one another. As the article states, "That pattern is the frozen echo of sound waves that propagated through the primordial soup of electrified gas and energetic photons filling all of space for the first few hundred thousand years after the Big Bang." While the original sound waves were much smaller in diameter than the huge structures that we see today—their current size is easily explainable and expected based on the expansion of the universe for the last 13.7 billion years.

9

The Standard Model of Particle Physics

Many people find the idea and concepts in modern particle physics to be very confusing. This is definitely understandable as there are literally hundreds of particles that are currently known in particle physics. Despite this, there is in fact a *standard model* for particles physics that provides clear-cut rules (as it were) for the different types of particles, and their fundamental properties. In this section, we will explore this standard model while attempting to keep confusion to a minimum. While researching this section of the book, I came across a remarkable web site that provides an excellent high-level overview of the standard model at: http://particleadventure.org/particleadventure. This internet document was put together by the Particle Data Group of the Lawrence Berkeley National Laboratory. It houses much useful information for all levels of readers; all in a format that is easy to navigate through, and has many useful illustrations. A good overview of the standard model in chart form can be found at the following web address: http://particleadventure.org/particleadventure/frameless/chart.html. This chart provides a good visual overview of all the fundamental groups and particles housed within the standard model's framework. This chart is a highly recommended visual reference to the text as the text descriptions of particles and interactions can quickly become overwhelming, as we will soon see. What is often confusing in a text medium can be more straightforward in a visual medium. The standard model of particle physics quite simply explains the leading theory of what the world around us is made of. With the discovery of quantum physics and the utilization of particle accelerators to probe the sub-atomic realm, it soon became clear that everyday matter was made up of smaller constituents, namely *atoms*. It should be noted that the standard model is a quantum theory in which all the key elements boil down to quantized increments of energy, mass, charge, and spin.

The Rutherford Experiment

Ernest Rutherford devised a way to test the theory of atoms in 1909 by construct-ing a famous experimental apparatus. In his apparatus, he had a radioactive source shielded in lead. In this lead shielding was a small opening where radioac-tive alpha particles (which are nothing more than helium nuclei composed of two protons and two neutrons) could escape in a specific direction. In the direction of travel of these alpha particles, Rutherford placed a thin sheet of gold foil. This sheet of gold foil was enclosed inside a screen coated with zinc sulfide. An open-ing in the screen was oriented such that the incoming stream of alpha particles could enter the screen (two dimensionally, this screen resembles the shape of the letter "**c**"). As an atom from the radioactive source decayed and emitted an alpha particle, that particle would travel into the screen chamber and interact with the atoms in the gold foil, thereby deflecting them by some calculable angle on their way to the zinc sulfide coated screen. Upon impacting the screen, the zinc sulfide would absorb the energy of the alpha particles and emit electromagnetic radiation in the form of a flash of light, thereby showing where on the screen the alpha par-ticle ended up.

The results of this experiment were startling. While most of the alpha particles went through the gold foil undisturbed, or nearly undisturbed, some of the parti-cles were deflected at huge angles. Some of these angles of deflection were so large as to deflect the alpha particles in the *opposite* direction they entered the chamber. This was a crucial step in realizing that the mass of atoms was concentrated into an extremely small volume of immense density. The following is an illustration of the Rutherford apparatus.

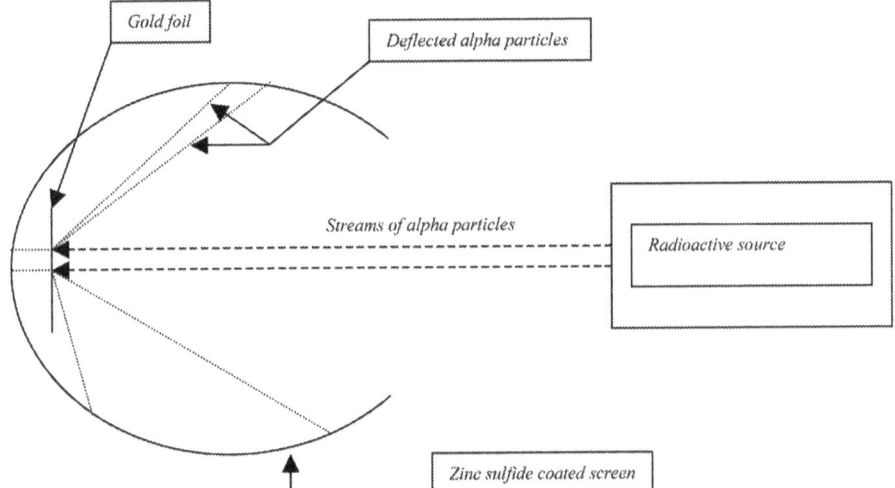

Illustration of the Rutherford experiment. A radioactive source emits alpha particles into a "C" shaped chamber with a piece of gold foil in the center. The alpha particles will pass through the foil without colliding with the gold atoms, or they will be deflected at various angles depending on the angle of incidence of the collision. After passing through the gold foil, or being deflected by the gold atoms, the impacts are recorded on the zinc sulfide lined chamber. Rutherford realized that in order for such huge angles of deflection to be accounted for, the mass of the atom must be concentrated into an extremely small region, and is not spread out uniformly throughout the atom.

Antimatter

All matter particles have a corresponding *antiparticle*, which is short for *anti-matter particle*. Antiparticles are identical to their matter-partners in every respect except for their electrical charge (for example, an electron has a charge of-1, while an anti-electron [*positron*] has a charge of +1). Anti-particles were predicted by the theorist Paul Dirac in 1928 when he was working with equations that described the behavior of electrons. In these equations, he would run into solutions involving square roots. As we know, if you take the regular square of any number, positive or negative, you end up with a positive number. Therefore, that same principle is why from math class we learn that square roots have both a positive and negative solution. For example if you take 8 and square it, you get 64. If you square-8, you also get 64; hence the solutions to the square root of 64 are 8 and-8 (you may also remember, perhaps unfavorably, the quadratic equation in algebra, which always has two solutions due to the square root term in the numerator). Dirac, instead of ignoring the negative solutions to the electron equations he was studying, proposed they might in fact represent a real particle—an anti-particle that accompanies the electron in all interactions. This proposal was largely ignored until 1932, when cosmic rays interacting with magnetic fields in a particle detector showed the trail of a particle that had the exact same properties of an electron, but was curving in the magnetic field in the opposite direction, which meant it had an exactly opposite charge to a typical electron. This detection of a positron (anti-electron) moved Dirac's theoretical proposal into physical reality. However this "reality" in which anti-particles live is very short; for as quickly as anti-matter is formed, it recombines with its matter partner and they annihilate into pure energy by Einstein's $E=mc^2$. One of the biggest questions that plaques the minds of physicists is why, if there is anti-matter particles for every matter particle, is there so much ordinary matter around? Where is all the anti-matter in the universe? Quite simply—we don't know. For reasons which are still being explored, there was slightly more matter that was produced than antimatter.

Quarks, Leptons, and Force Particles

In our current understanding of particle physics as represented in the Standard Model, all of the world around us can be described in terms of six quarks, six antiquarks, six leptons, six antileptons, and force carrying particles—that is it. It should be noted that the force of gravity is *not* included in the standard model of particle physics. Due to the weakness of gravity in proportion to the other forces of electromagnetism, and the strong and weak nuclear interactions, it is currently unable to be housed under the standard theory framework. Do date, all of the particles predicted by the standard model have been verified. Standard model agrees to experiment to an extraordinarily high degree of precision.

Quarks

Particle physics has been unraveling layer upon layer of "fundamental" particles. For instance, the atom used to be considered the underlying components of matter. As time passed, and experiments and theories expanded the depths to which they could probe, this turned out to be false as the atoms were found to be composed of electrons, neutrons and protons. Quarks represent the next layer of underlying components beyond protons and neutrons (as far as we know—electrons are still fundamental). Unlike an electron that has an integer charge of-1, quarks have fractional charges based of either +/-2/3 or +/-1/3. Quarks are never found in solitude; rather they always are in pairs or triplets whose combined charge adds up to an integer amount. Quarks are the fundamental components (unless string theories prove correct, or higher powered accelerators unpeel another layer of the particle "zoo") of all the particles in the baryon and meson classes. Baryons are simply particles that are made up of three quarks, such as protons and neutrons. Mesons are particles that are made of one quark, and one anti-quark. Baryons and Mesons are part of a larger group called Hadrons. A hadron is any particle whose net electrical charge equals to an integer value.

To make things even more confusing, the six different quarks are identified by different names that are referred to as *flavors* in particle physics. So adding in the flavors gives us "up" quarks, "down" quarks, "top" quarks, "bottom" quarks, "strange" quarks, and "charmed" quarks, for our total of six. Quarks also all have a *spin* of ½. This concept of spin is not to be taken literally and is a confusing reference. In experiments, particles seem to have miniscule *magnetic moments* associated with them (just as our spinning earth generates a magnetic field), and their own angular momentum, so the term *spin* was applied to them. In reality, they are not literally "spinning." Particle spins can be either zero, or increments of 1/2 (e.g. 0, ½, 1, 3/2, 2) up to two. Quarks all have a spin of ½, and have unique masses.

The protons and neutrons of atomic nuclei are made of three quarks each. A proton is the combination of two up quarks and a down quark (*uud*), and a neutron is the combination of one up quark and two down quarks (*udd*). Up quarks have a charge of 2/3 and down quarks have a charge of-1/3, therefore if we add up charge the constituents of a proton's up-up-down quarks, we get 2/3+2/3-1/3, which equals +1. For a neutron, made of up-down-down quarks, we get 2/3-1/3-1/3, which equals zero (or *neutral*) charge. All throughout the standard model of particle physics, we see Einstein's imprint in the form of mass-energy equivalence as presented by $E=mc^2$. For instance, if you add up the masses of the up and down quarks that compose protons and neutrons, they do not add up to the mass of a proton or neutron. This is because most of the mass of protons and neutrons (and other particles) arise from their kinetic and potential energies; and since mass and energy are equivalent by $E=mc^2$, this kinetic and potential energy shows up as a deficit between the combined masses of the quarks, and the total mass of protons and neutrons, thereby maintaining the law of conservation of mass and energy. Quarks also have another charge type as well. This attribute is referred to as *color*. We will explore this concept of color further later in this section. A quark's color determines how it will interact with other quarks and be affected by the strong force.

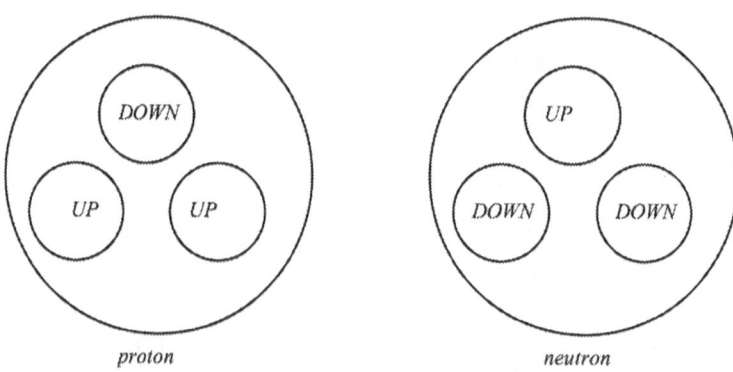

proton   neutron

Quark constituents of protons and neutrons. Protons are comprised of two up quarks and one down quark, while a neutron is comprised of two down quarks and one up quark.

Leptons

Another class of subatomic particles is the lepton. Despite lepton being Greek for "small mass", some leptons are actually quite massive in comparison to other particles. Unlike baryons and mesons that are composed of quarks, leptons are solitary particles. Our familiar electron is a member of the lepton group. All the particles in the lepton group are as follows: the electron neutrino, electron, muon neutrino, muon, tau neutrino, and the tau. The muon and tau particles share the same charge as the electron, however are much more massive. *Neutrinos* are nearly massless, have no electrical charge, and rarely interact with matter. Neutrinos are so elusive, that they periodically pass through the entire earth without interacting with a single atom along the way (a neutrino could pass through light-years of lead and not interact with anything as well). Neutrinos existence was first put forth to explain the apparent missing momentum observed when radioactive nuclei decay (such decay is a primary source of neutrinos, along with nuclear fission/fusion). As experiments progressed, the neutrinos existence was detected in particle accelerators. Despite their tiny mass, neutrinos are theorized to have been produced in huge quantities during the first stages of our cosmos. In fact, they could have been produced in enough quantities to direct the mass and expansion of our present cosmos.

Generations of Matter

Moving along in the standard model's particle zoo, we come to the *generations* of matter. Both quarks and leptons are composed of three such generations, with each generation consisting of four particles. Essentially, the generations of matter represent the stages of particle decay sequences as a function of mass. Generation III particles are the most massive, followed by the less massive generation II, and the least massive generation I particles. All of our ordinary matter that we are familiar with is composed of generation I particles. Generation II and III particles are unstable, and quickly decay into generation I particles (a big unanswered question in particle physics is why there are generation II and III particles that exist at all). The breakdown of the particles in the generations is as follows. In generation I, you have the up quark, the down quark, the electron neutrino, and the electron. Generation II consists of the charmed quark, the strange quark, the muon neutrino, and the muon. And finally, for generation III, we have the top quark, the bottom quark, the tau neutrino, and the tau.

The Forces of Nature under the Standard Model

Under the umbrella of the standard model, we are able to explore the interactions of three of the four fundamental forces of nature: electromagnetism, the weak interaction, and the strong interaction. As we stated previously, gravity is not able to be housed under standard theory. Attempts are underway to incorporate this, and it is hoped that some day it will be included, thereby making a theory of everything of all the forces. The *graviton*, the proposed force carrying particles of the gravitational force has not been detected yet. This is not surprising due the extreme weakness of gravity and the lack of interactions that arises from it on quantum scales.

My father once said that he would never be able to understand how gravity worked. More specifically, he did not understand how a force could act between objects when there was no tangible body between them. An excellent analogy that helps clear this up (and hopefully for my father as well) at least in regards to repulsive forces is two visualize two people facing each other on an ice rink, wearing ice skates. Now, let us imagine giving one of them a brick to toss back and forth between each other—to play "catch" with. What happens? Since the ice has so little friction, and the brick has mass, as they swing their arms to give the brick enough energy to traverse the distance to the other person, they get pushed back. And, as the other person catches the brick, they absorb with their body the energy that the first person put into the brick to get it there, so they also move back. Each time the brick is thrown or caught, the people move away from each other on the ice. If you made the brick such that the people on the ice could see it, but everyone else could not, then it would *appear* that the ice skaters were being repelled from each other by an unseen force. In reality, the "brick" is a *messenger particle* that carries the force with it (even after using this example, I must admit that I agree with my father, in that there will always be a certain mystical quality to forces such as gravity—and I prefer to keep it that way). While this is an analogy of a repulsive force, it still provides a mental image to visualize.

All of the messenger particles for the forces of nature are rolled up into their own class called *bosons*. Bosons have integer spins of 0, 1, and 2. For the electromagnetic force, the force particle is the familiar photon (γ). For the weak force, there are three force particles; they are the W^-, W^+, and Z^0 bosons. In the strong force, gluons are the force carrying particles. All of these bosons are akin to the different types of "bricks" would could give our people on the ice rink to toss back and force between each other. Bosons have certain underlying rules that govern which particles will be able to "feel" the force they are carrying (e.g. electrons and protons have an electric charge, so they absorb and produce photons, the carrier particles of the electromagnetic force).

The Electromagnetic and the Residual Electromagnetic Force

The electromagnetic force was discovered by James Clerk Maxwell. From his research and insight, he found that he could combine the once separate magnetic and electrical forces under a single force by the use of four equations (presented in the form based on the absence of magnetic or polarized media):

$$\oint \vec{E} \cdot d\vec{A} = \frac{q}{\varepsilon_0}$$

$$\oint \vec{B} \cdot d\vec{A} = 0$$

$$\oint \vec{E} \cdot d\vec{s} = -\frac{d\Phi_B}{dt}$$

$$\oint \vec{B} \cdot d\vec{s} = \mu_0 i + \frac{1}{c^2} \frac{\partial}{\partial t} \int \vec{E} \cdot d\vec{A}$$

The specifics of each term are not important for our discussion here; however I feel it is important to at the very least have a visual representation of these famous equations as they will be frequently referenced should you be interested in further readings on the subject of electromagnetism. From his equations describing electromagnetism, Maxwell found out that when he calculated the velocity of these waves, he always came out with the same answer—the speed of light. He therefore correctly concluded that all photons must be nothing more than electromagnetic waves. This speed is independent of wavelength, and therefore all of the radio, microwave, infrared, visible, ultraviolet, x-rays and gamma rays that make up the electromagnetic spectrum, all travel at the speed of light. Additionally, all types of electromagnetic radiation are composed of photons, as they are the force carrying boson for this force as stated in the standard model. Interestingly, Maxwell originally started working with hydrodynamic equations to get him started on his unification of magnetic and electrical forces. He envisioned the positive charges of the electrical field as being water "sources" and negative charges as being the equivalent to "drains". This approach, with the concept of a *force field* put forth previously by Michael Faraday, allowed Maxwell to unify electricity and magnetism in such an elegant manor.

As we know, the electromagnetic force is what causes the repulsion or attraction of objects that have electrical charges, where like charges repel, and opposite charges attract. The basic rules that govern the symbiotic relationship between electricity and magnetism are as follows: charges produce electric fields, *moving* charges produce magnetic fields, *changing* magnetic fields produce electric fields, and *changing* electric fields produce magnetic fields. This relationship between electric and magnetic forces is readily apparent to anyone who has tried to use a compass outside when a lightning storm is nearby. The changing electric fields of the thunderstorm produce their own magnetic fields that interfere with the earth's magnetic field, there by causing a compass to move erratically. Just like gravity, the strength of the electromagnetic force drops off with the *square* of the distance from the source of the radiation. Despite this relationship, the distances that electromagnetic and gravitational forces can effect objects is far greater as compared to the weak and strong forces, which are bound to sub-atomic scales.

Atoms tend to be electrically neutral, with the number of electrons with negative charge equally the number of protons with positive charge. This balance is what gives atoms their stability (if this is out of balance, then radioactive decay takes place). Groups of atoms are attracted together to form molecules by what is referred to as the *residual electromagnetic force*. This arises from the fact that electrons in nearby atoms of matter are attracted to the positive protons in their atomic neighbors, and vice versa. This attraction forces certain orientations of atoms, and is the production mechanism of molecules. However, this residual electromagnetic force comes into balance once atoms get a certain distance apart due to the electromagnetic repulsion of each atom's outer electron clouds, which contains like charges that repel one another. What we observe as "chemistry" comes about from these electromagnetic interactions.

The Strong Interaction

In order to keep the protons and neutrons together (and their constituent quarks), there needs to be a force much more powerful than the electromagnetic force. This interaction is the *strong interaction*, which bonds together the atomic nucleus. The strength of the strong force must overcome the electromagnetic forces within atomic nuclei in order to keep the positively charged protons from flying apart due to their repulsion. The quantum theory that explains the strong interaction within the standard model is referred as *quantum chromodynamics*. As we can remember from earlier, all quarks that make up baryons have electrical charge. They also have another type of charge that is called a *color charge*. There are three color charges for quarks (red, green, blue) and three color charges for antiquarks (anti-red, anti-green, anti-blue) This color charge is what governs the strong interaction. The carrier particle of the strong force is the *gluon* (as they "glue" the nuclei of atoms together), which also has a color charge (gluons carry both color charge and an anti-color charge). Despite having a color charge, when quarks combine to form a baryon, that combination cancels out the color to a neutral charge. In this way, color charge is a conserved quantity. What this means, is that the strong force interactions take place between the quarks themselves, on incredibly small scales compare to the electromagnetic force. When the electromagnetic force is at work, photons are exchanged, when the strong interaction is at work, gluons are exchanged. Since the gluons of the strong force are only exchanged at such small distances inside the atomic nucleus, there are some peculiar features that it must have in order to maintain stability. One of these oddities is that the strong force becomes stronger by exchanging more and more gluons as quarks are moved apart. A good analogy for the strong interaction is to imagine a rubber band around your fingers. If you keep your fingers relaxed, you still feel the force of the rubber band, but not nearly as much as when you start separating your fingers and stretching out the rubber band. The more you stretch out your fingers, the more force the rubber band exerts to keep your fingers in place. If your fingers are strong enough, you may eventually be able to stretch the rubber band to a point where it suddenly breaks all together. This same thing happens when color force fields are being exchanged via gluons. If enough energy is put into the quarks, eventually, the gluons are unable to "glue" them together any longer, and the extra energy of the gluons is used to create a new quark-anti-quark pair, thereby conserving the energy of the system.

Just as the electromagnetic force has a residual electromagnetic force associated with it, so too does the strong interaction have an accompanying *residual strong interaction*. In the strong interaction, the residual interaction represents the attractions between the quarks in the protons of a nucleus. This residual strong interaction is what actually binds the nucleus together and gives it stability by overwhelming the repulsion of the like charges of the protons.

Electroweak Force and the Weak Interactions

When particles decay to lighter particles, this is done via the *weak interaction*. As you will remember, the generation I particles are what we see around us as they are not able to decay any further and are thus stable. The carrier particles for the weak interactions are the W^+, W^-, and Z particles, with the W^+ and W^- having electrical charge, while the Z is electrically neutral (sometimes denoted as Z^0). When a decay occurs, the total mass and energy must be conserved, however some of the original particles mass gets converted into kinetic energy—again, via Einstein's $E=mc^2$. The remaining particles after decay has occurred always have *less* mass than the particle it decayed from as a result of the expended energy (which corresponds to a loss in mass).

The standard model unites the electromagnetic and weak interaction into a unified force that is referred to as the *electroweak interaction*. This is possible because at the extremely short distances involved in sub-atomic interactions, the magnitudes of these two forces are comparable. However, at greater distances, the magnitude of the weak interactions becomes very small, where the electromagnetic force remains much stronger. The strength of the force that is generated by the electroweak force is dependent on the mass of the force carrier particle, and the distances at which the interaction is taking place. Since the W and Z carrier particles of the weak force have such an immense mass compared to the massless photon carrier particle of the electromagnetic, the range of distance that the electromagnetic force is able to be "felt" is much larger than for the weak interaction.

In the early universe, when conditions were much more dense and hot than we observe today, the electromagnetic and the electroweak forces were united. A good description of this, which I will present here is in John Gribbin's book The Search for Superstrings, Symmetry, and the Theory of Everything (page 133): "When the energy density (temperature) of the Universe was great enough, particles...could appear spontaneously, in particle-antiparticle pairs. And instead of a carrier of the weak interaction being able to come into existence only for the brief instant of time allowed by the uncertainty principle, the supply of free energy around it could make any of these virtual particles real, and give it an extended lifetime. As long as the mass of the particle was less than the energy available, it could live forever, like the photon, and the distinction between W's and Z's would be dissolved away. At high enough energies, during the early phases of the Big Bang, there is no distinction between the electromagnetic and the weak forces. The distinction only appears because we live in a cold Universe, where the symmetry is broken. The W's began to freeze out of our Universe when the temperature fell to 10^{15} K, about one thousand-millionth of a second after the moment of creation."

W and Z particles are responsible for radioactive decay. There are three types of decay that are possible in the current standard model: alpha (α) particles, beta (β) particles, and gamma rays (γ). Alpha particles are positively charged and are helium nuclei that have been stripped of their surrounding electrons. Beta particles are electrons, and are thus negatively charged. Gamma particles are electrically neutral and represent the most penetrating of the decay products (which is why they are harmful to human tissues). A page of this book will stop an alpha particle dead in its tracks. To stop a beta particle, you would need a sheet of aluminum foil. For a gamma ray, you would need a block of lead. On a side note, you can determine what type of radioactive decay is occurring by having the alpha, beta, or gamma rays pass through a magnetic field. Alpha and beta particles will curve in a particular direction according to their electrical charges and the orientation of the magnetic field, while gamma rays, will not be affected as they are electrically neutral. Radioactivity was first discovered by Henri Becquerel. When photographic plates were placed in the vicinity of certain metallic ores, such as uranium, he noticed that the photographic plates would become exposed, even if he covered them in dark paper. He correctly concluded that there must be energetic particles that were spontaneously being emitted from these ores without any external energy input.

We now know that radioactivity is caused from unstable atomic nuclei. This instability is caused by an inequality in the number of protons and the number of neutrons, thereby setting up an imbalance. As time goes on, substances made of radioactive material will spontaneously decay over time until they have stabilized the imbalance of protons and neutrons. This spontaneous emission of radiation over time gives rise to a substances *half life*. The half-life of a radioactive material is exactly as it sounds—it is the time it takes for half of a given amount of matter to decay into a stable form. While the event of a particular radioactive decay is governed strictly by quantum probabilities, over time, these decays give rise to a very predictable pattern that can be utilized to accurately determine the age of certain objects (carbon dating methods use this principle). Again, in this radioactive decay, Einstein's mass-energy equivalence plays a crucial role. For instance, as uranium decays into thorium, alpha particles are released. In accordance to Einstein's $E=mc^2$, some of the mass of the original uranium must be converted into kinetic energy for the emission of the alpha particle. This conservation of energy shows up as a loss of mass when comparing the mass of the uranium, minus the mass of all the particles it produces while decaying.

Quantum tunneling turns out to be the culprit for how radioactive decay occurs. According to the uncertainty principle we cannot know exactly the location and momentum of a particle at the same time. If we measure the location more precisely, then we must pay a price for this increased accuracy in a less-precise measurement of momentum, and vice versa. Remember that in quantum tunneling, there is a possibility that on the smallest of time scales (Planck time) these particles can "tunnel" through an energy barrier and suddenly find themselves outside an atomic nucleus. For smaller nuclei with few protons and neutrons, the probability of this is incredible small (but possible—for instance a proton is thought to decay, but over an unimaginably long time). However, for the largest of atomic nuclei, such as uranium nuclei, the odds of quantum tunneling occurring are greatly increased, and every now and again, a set of two protons and two neutrons (an alpha particle) will suddenly tunnel their way outside the residual strong force, and escape—by seeming spontaneous means. We see this as spontaneous radioactive alpha decays.

It is important to realize that the electromagnetic force, the weak interaction and the strong interaction *all* result in different types of particle decays. The main difference to keep clear is that the weak interaction is the only way that *fundamental* particles can decay.

The Range of the Forces

The range of gravity and the electromagnetic forces are infinite. One atom placed at one end of the universe will be able to feel the force of another atom placed at the other the of the universe with these forces (that is, of course, if enough time has pass for the force to travel between them at the speed of light—we must not forget Einstein). A photon emitted by the electromagnetic force will be able to travel across the whole universe. This is why we are able to detect the faint photons of light from distance galaxies at the edge of the universe. For the weak interaction, this is not the case. The ranges of which the weak interaction can be felt are extremely small as compared to gravity and electromagnetism (about a 10^{-18} meters, or about 1000 times smaller than the diameter of an atomic nucleus). Why is this?

The first step in understanding this is to look at the carrier particles of each of these forces. For gravity, the proposed graviton is massless; in electromagnetic waves the photon carrier particle is also massless. In the weak interaction, the W and Z bosons are the carrier particles, and they have masses of 80.4 and 91.187 GeV/c^2, respectively. This brings us to the second step, which is the invocation of the uncertainty principle. As we will remember from earlier, the uncertainty principle can be written in terms of energy E and time t as follows:

$$\Delta E \Delta t \geq \frac{\hbar}{2} \text{ or } \Delta(mc^2)\Delta t \geq \frac{\hbar}{2}$$

Following this principle, and remembering the equivalence of mass and energy via $E=mc^2$ we can see that if we have a massless carrier particle such as a graviton or photon, then the uncertainty in time, or the *lifetime* of the particle as allowed by this equation is infinite. Conversely, if we plug in a particle with a non-zero mass such as the W and Z bosons of the weak interaction, we see that as the energy (mass) of the carrier particle increases, the time t of the principle must become smaller and smaller, thereby decreasing the lifetime allocated to such a particle. To reiterate, part of the key to the establishment of this uncertainty principle is based on light being the fastest that information can be carried through space-time.

A more succinct description of this can be found in John Gribbon's book, The Search for Superstrings, Symmetry, and the Theory of Everything (pages 71-72): "In electromagnetic field theory, the force results from the exchange of particles, virtual photons. Because photons have zero mass, the amount of energy a photon carries can be made vanishingly small by giving it a very long wavelength. So there is not limit, in principle, to the range over which electric [and magnetic] forces can be felt—a virtual photon associated with an electron can interact, albeit very weakly and with very low energy, with another electron anywhere in the Universe—although, of course, the interaction is much stronger if the electrons are close together...But what if photons had mass? In that case, there would be a certain minimum amount of energy that would be required in order to make a virtual photon, E. And the finite size of that packet of energy would set a firm time limit t, on the life of such a particle, in line with Heisenberg's uncertainty relation. Since nothing can travel faster than light, this finite lifetime would mean that any such particle...would have only a very limited range, since it has to return to its origin, or find another particle to bury itself in, before its allowed lifetime is up." For a W and Z boson of the weak interaction, this lifetime corresponds to around 3×10^{-25} seconds, which in turn corresponds to the range of 10^{-18} meters that the weak interaction acts over when traveling at the speed of light.

What about the strong interaction? The carrier particle of the strong interaction is the gluon. Gluons are massless. Despite this, the strong interaction acts over an extremely limited range. This is because unlike the other forces of gravity and electromagnetism, and the weak interaction, *the strong interaction also affects the carrier particles themselves.* The gluons that are exchanged during strong interactions have associated color charges which "feel" the strong interaction. This leads to the limited range that the strong interaction can be felt despite the fact of its gluon carrier particles being massless like the photon or graviton.

Fermions, Bosons, and the Pauli Exclusion Principle

Regrettably, if the outline of the standard model of particle physics above has proven very confusing, unfortunately, we have a couple more items to cover before we are out of the thicket. For certain particles are not allowed to exist in the same quantum state at the same time—this phenomenon is known as the *Pauli Exclusion Principle*. The particles that have restricted quantum states in accordance to this principle are known as *fermions*. Particles that are not bound by the Pauli Exclusion Principle are referred to as *bosons*. Fermions have fractional spins, more specifically they have half-integer spins (such as $\frac{1}{2}$, $3/2$...etc). Quarks, protons, neutrons, and leptons are some examples of fermions. Bosons, on the other hand, only have integer spins (0,1,2...etc). All of the force carrying particles fall into the boson family (including the theorized, but not yet observed spin 2 graviton). For atomic nuclei, if it is made up of an odd number of protons and neutrons, then it is a fermion. Conversely, if atomic nuclei are made up of an even number of protons and neutrons, then it is a boson.

Symmetry and Supersymmetry

I must admit to a bit of carelessness on my part in not introducing symmetry and supersymmetry earlier in this book. However, it is as easily incorporated here as is would be in many other sections. There are many readily available examples of symmetry in our everyday experiences. For instance, if you take a marble that is all one color, and rotate it, it appears the same. In fact, you can rotate it in any possible way and it will still appear the same. This spherical marble has a high degree of symmetry. Let us now consider another object, a circular drinking glass. If you were to set this glass on a table as if it to fill it up with water, you can obviously see that if you rotate the glass through an axis that runs perpendicular along the center of its base, it appears visually unchanged. If however, you place the glass on its side and rotate it, the appearance of the glass changes as rotations occur. The circular drinking glass has *rotational symmetry* about one axis only. When you place the glass on its side, you have *broken* its symmetry. Physicists often refer to symmetry breaking—and it is merely a transformation that no longer allows for a certain type of symmetry to take place anymore, just like placing our circular glass on its side.

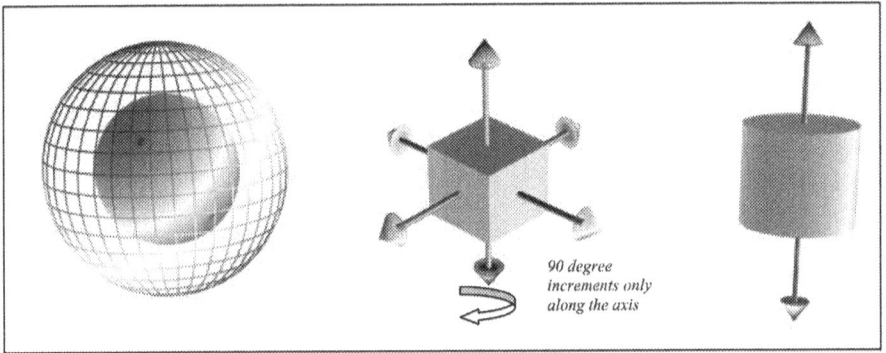

Three examples of symmetry from high symmetry to lower symmetry. A sphere is able to rotate along any point the grid in the figure that passes through its center without changing appearance—it will remained unchanged under those rotational transformations. A cube can be rotated in 90 degree increments along axis that run perpendicular to the center of each face. A cylinder can only be rotated through the axis that is perpendicular to the centers at the ends of the cylinder.

The term *supersymmetry* has its roots in particles physics, and is often referred to by the acronym *SUSY*. In particles physics, there is a theorized symmetry among all of the matter and force particles (more specifically fermions and bosons). Supersymmetry requires that for every matter particle, there must be an accompanying supersymmetric force particle, and for every force particle, there is an accompanying supersymmetric matter particle. These proposed accompanying supersymmetric particles are called *superpartners*. Such superpartners have very odd names. For instance, the superpartners of quarks are called *squarks*, and the superpartners of neutrinos are *neutralinos*. To date, there has been no experimental evidence for the existence of supersymmetry. The energy levels required to create the superpartner particles that are theorized to exist could be detect in the next round of high-energy colliders such as the Large Hadron Collider. A good overview and design of several major accelerators is outlined by the <u>Lawrence Berkeley National Laboratory</u>, at: <u>http://particleadventure.org/particleadventure/frameless/variation.html</u>.

The Mass of Particles and the Higgs Field

One of the problems associated with the standard model is that is does not account for how particles acquired their associated masses. Unfortunately, the standard model requires that we provide it with the particle masses. This is another reason why physicists believe that there must be a deeper theory such as M-theory that will not require particle masses to be inserted, but rather that the masses would naturally emerge from the framework of the theory itself. Peter Higgs, in 1964 proposed a new concept that is know as *spontaneous symmetry breaking*. We can use our drinking glass to show an analogy of this spontaneous symmetry breaking. As we remember, our glass has rotational symmetry about one its axis's. If, however, we have our glass sitting on the table on its symmetrical axis, and then pour boiling hot water into it—what happens? The sudden change in energy (temperature) causes the glass to shatter, thereby *breaking* the previous symmetry. For Peter Higgs, he mathematically represented this by introducing a new *field* which would give rise to particles observed masses. This *Higgs field* is theorized to permeate all of space. At high enough energies, such as what were present in the early instants of the cosmos, this field would have been perfectly symmetric (think of our marble). As the universe cooled, this symmetry suddenly broke, and in accordance to Einstein's equivalence of matter and energy, the energy that was pent up in the symmetry transforms itself into mass (much like a top spinning about its axis of rotational symmetry contains inertia—so that when you suddenly stop the top from spinning, you transfer that inertial energy into your hand). Whatever particles were present during the spontaneous symmetry breaking of the Higgs field will acquire mass from the energy of the field.

10

String Theory—Super String Theory—M-Theory, Overview

Quantum mechanics and relativity are fantastically successful theories. Quantum physics and relativity provide what are widely referred to as the twin pillars of modern physics. While our everyday intuitions consistently lead us astray when we try to apply them to these fields, they have both been experimentally proven to extraordinary accuracy, and we therefore must let those results speak for themselves. However, there is a problem that has plagued physicists for decades. We have two successful theories, one which is good at accurately describing large scale systems where copious amounts of matter bend space-time; that of relativity, and another theory that accuracy describes what is happening on the smallest of scales between elementary particles, that of quantum physics. The problem is that these two theories are incompatible mathematically. When you try to combine the non-probability based equations of relativity with the probability based equations of quantum theory, you get non-sensible answers to problems, such as infinities, negative probabilities, or probabilities greater than one hundred percent. What this means is that these two theories cannot be unified. There needs to be another type of theory and associated mathematics that must be explored in order to unify relativity and quantum theory. The theory that is generating the most excitement and that has the strongest potential for the unification these two theories is *string theory*. The basic principle behind string theory is quite simple: all the observed properties that we see in the cosmos can be boiled down to the interaction of *tiny vibrating strings of energy*. These strings are of incredibly tiny if they exist, being on the order of a Planck length, or around 10^{-35} meters in length. This tiny size corresponds to about 10^{-20} times the diameter of a proton. In other words, it would take 10^{20}, individual string diameters to equal the length of a single proton. The problem with the incredibly small distances that strings occupy is that we have little to no hope of detecting them directly through experiments. Another way of picturing of how small these strings are according to this theory is outlined in Paul Halpern's book <u>The Great Beyond</u> on page 251: "This scale [of strings] is so miniscule that if a hydrogen atom were blown up to the proportions of the Milky Way galaxy, the strings within it would only be the size of dust mites." The figure below outlines the breakdown of matter into its constituents according to string theory.

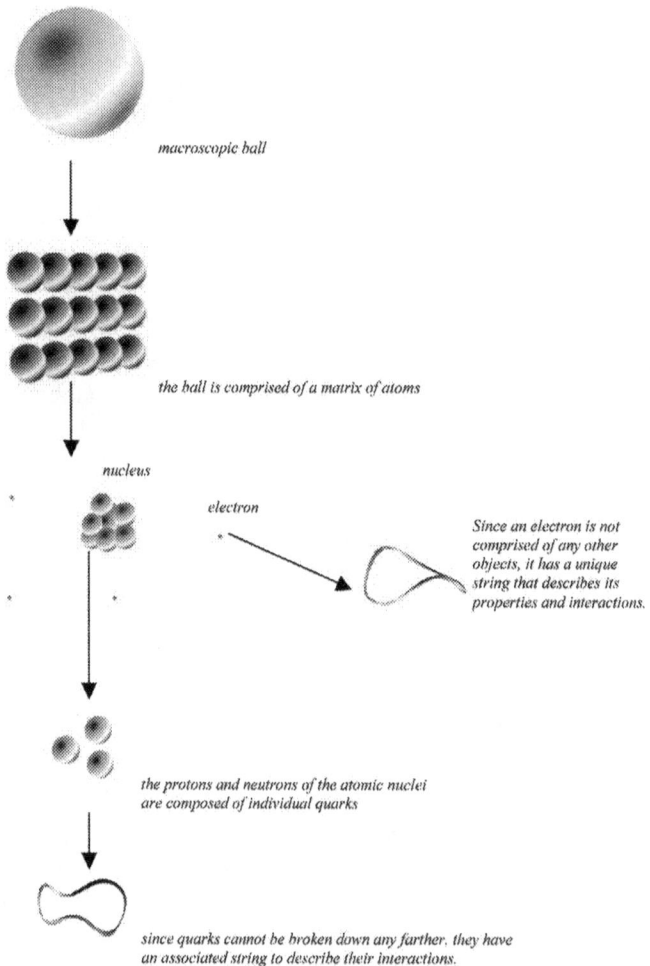

macroscopic ball

the ball is comprised of a matrix of atoms

nucleus

electron

Since an electron is not comprised of any other objects, it has a unique string that describes its properties and interactions.

the protons and neutrons of the atomic nuclei are composed of individual quarks

since quarks cannot be broken down any farther, they have an associated string to describe their interactions.

A visualization of the constituents of matter according to string theory by examination of a macroscopic ball at various magnifications. The ball is composed of arrays of atoms, whose electrons have their own accompanying string. The protons and neutrons of the atomic nuclei are composed of individual up and down quarks. In turn, each of these quarks has a unique quark string associated with them. In string theory, every fundamental particle has its own constituent string associated with it.

Why is this unification between relativity and quantum theory so important to physicists anyways? Physicists are always trying to describe the largest variety of properties with the fewest number of underlying rules. Einstein's relativity provides a detailed picture of gravity, space-time, near-luminal velocities, and how these behaviors are affected based on different frames of reference. Quantum physics provides a unified theory of all physical phenomena *except gravity* (the electromagnetic forces, the nuclear strong force, and the nuclear weak force). If we find a way to combine these two theories, we have achieved a *theory of everything*.

The reason that this unification is so important is that in the beginning of the cosmos during the big bang, there were massive densities that needed relativity to accuracy explain them, squeezed into a small enough size that requires quantum physics to described them as well. Additionally and more immediately, *black holes* require us to invoke both quantum physics and relativity, as they are immensely massive objects whose gravitational field has bent the surrounding space-time to such a degree that light itself cannot escape, and all this is happening in the tiny *quantum sized singularity* at the black-hole's center. So following this line of thought we strongly feel that there was a point in the earliest moments of the cosmos when all of the forces of nature were unified into an individual force ("the force" for all of you Star Wars fans). The discovery of such a theory of everything would be one of the greatest achievements of mankind, and we may be on the brink of it with the ongoing development of string theory. Due to the vast amount of configurations that are available mathematically by representing the smallest component of a quantum object as being built from vibrating strings of energy, string theories are successfully able to house relativity and quantum mechanics under the same roof....with some controversial requirements which we will cover shortly. For the interested reader, there is an excellent web site: http://superstringtheory.com which is dedicated to providing the latest information on developments on this rapidly changing theoretical area.

The problem with Gravity

In his later years, Einstein grappled with trying to unify the ideas of his special and general relativity theories with the other then known forces of nature: the electromagnetic force (the strong interaction that glues the atomic nucleus together and the weak interaction that produces radioactive decay were not known until later). Einstein never could find peace with the fundamental probabilities that quantum mechanics dictated must exist in every system, hence his often quoted line "god does not play dice". Ironically, Einstein's discovery of the quantization of light via the photon from his research on the photoelectric effect helped provide the foundation for the quantum theory which he disagreed with so deeply. Despite his immense efforts, he was unable to unify all the forces of nature in a theory of everything before he died on April 18th, 1955.

With string theories and M-theory (which we will cover shortly), some 50 years later, physicists are tantalizingly close to achieving Einstein's dream, albeit with a very different theory than he would have envisioned. What is the reason behind gravity's reluctance to unification? What makes it so much different than the other forces of nature? The answer is strength. Gravity is profoundly weaker than the other forces of nature. The electromagnetic force, which is much weaker than the strong nuclear force that provides the glue to hold atomic nuclei, is 10^{38} times the strength of the gravitational force. That is a truly immense difference. People often think of gravity as being a powerful force, as it is the prominent force that we have experience with. What you must realize is that it takes trillions and trillions of atoms gathering together for us to even feel the effects of it. Conversely, if we took the repulsive forces between electrons as represented by the electromagnetic force, and expanded them to the size of objects that gravity needs to be felt, then it would take *millions of tons* of force to stop those electrons from moving away from each other. Another way to look at it is this: it takes the entire mass of the earth, all 6000 *billion billion tons* of it, to generate the gravitational field that we feel everyday. Despite this immense bulk of matter, we can easily pick up and move many everyday objects and overcome this gravity. Even a small magnet can overcome the entire earth's gravity to lift objects. This immense difference in strength is what makes it so difficult mathematically to house gravity under a unified theory with the other three forces of nature; however it is not the only hurdle. Both quantum physics and relativity are supposed to be valid everywhere in the cosmos, however they do not agree when combined, and there cannot be two "everywheres" in *one* cosmos—thus making their unification inevitable in a theory of everything.

Kaluza-Klein Theories

While the notion of higher dimensions had been known for some time prior to quantum theory, the work of Theodor Kaluza and Oskar Klein is what is commonly referred to as the birth of modern string theories. Kaluza, in 1919 developed a theory that provided unification of Einstein's general relativity and

Maxwell's electromagnetic theory. To accomplish this, he combined Einstein's and Maxwell's equations into a *five dimensional* mathematical framework. There had been several other unification theories that were presented previous to Kaluza's theory, however what set Kaluza's model apart was that it left the core principles of general relativity unchanged (contrary to unification theories presented, for instance, by Nordström and Weyl). With the extra degree of freedom in his five dimensional space, Kaluza found that with a single set of relationships he could house electromagnetism and general relativity in his unification theory.

Oskar Klein had a similar discovery to Kaluza, however in a different context. Klein officially developed the first five dimensional theory that explained quantum phenomenon (with everything coming in discrete "packets" instead of a continuum that is non-discrete). Klein received a significant portion of his ideas from work that was done previously by De Broglie (the discoverer of matter waves). De Broglie envisioned electrons around an atomic nucleus being described by *standing waves* on a circle around the nucleus. Klein knew this could be a key to mathematically explaining the quantization of electrical charges. Just as a standing wave on a violin, guitar, or cello are restricted to the tones that are played based on finger placements, the standing waves setup around an atomic nucleus can only have certain frequencies, and therefore become *quantized* just as they are in quantum theory. Klein was taken by this idea, and imagined electrons as being on the crest of these standing waves, moving around with them as a surfer draws energy of motion from the wave they are riding. Klein used De Broglie's standing wave proposal in his theory and found that in his five dimensional framework with standing waves, there could be only whole numbers of waves around the atomic nucleus. Additionally, Klein showed that a particle's momentum is *inversely* proportional to its wavelength. This inverse proportionality enabled Klein to be able to calculate the size of the fifth dimension in his model. This turned out to be 10^{-30} inches. This was comforting to Klein, as in reality there is not any evidence for higher dimensions beyond our three of space, and one of time. However with such a tiny size, Klein showed that this would make sense if a higher dimension's presence was hidden due to its miniscule size.

The efforts of Kaluza and Klein have often been used in subsequent developments of higher dimensional theories. Theories that have come forth based on their ideas, or that are direct derivatives of their ideas are known collectively as *Kaluza-Klein theories*. Our current string, superstring, and M-theory are derivatives of Kaluza-Klein theories. Much of Kaluza and Klein's work, along with the work of Weyl, Nordström, Reinmann, Minkowski, and Fock (to name a few) paved the way for exploring the mathematics, and the potential unification powers of higher dimensional spaces.

Unification powers

In particle physics, subatomic particles are represented by point-like objects that are one dimensional. In light of our quantum view, we need to consider them as "shells of probability" surrounding an idealized point. So what is it that provides string theory with such a unifying power that can combine quantum theory and relativity under one roof? The key lies in the degrees of freedom associated with their hypothesized core piece; that of vibrating strings of energy. Just as any stringed musical instrument is capable of a wide range of tones based on where the musician has their fingers placed, the constituent string of energy in string theories allow a similarly wide range of mathematical variations. With point particles, there are only a certain number of ways that you can orient or move around an object that is theorized to be a particle; you can jiggle it around and "spin" it (spin is what gives particles their electrical charge), however that is about it.

A vibrating string of energy has vastly more ways in which it can vibrate, move, and change over time. Mathematically this affords them a great flexibility in being able to accommodate the tiny strengths of gravity along side the comparatively massive electromagnetic, weak, and strong forces of nature. In fact, there are so many degrees of freedom required of a string, that modern superstring theory (which stands for supersymmetric-string) and M-theory (the "M" stands for matrix, mysterious, or magic—depending on your preference or further experimental results) require 9 and 10 spatial dimensional spaces (respectively) in order for the theories to be viable. The term "viable" here means that in order to explain all of the phenomena housed within the forces of nature, superstring and M-theory required 9 and 10 spatial dimensions to provide ample *degrees of freedom* for the strings to vibrate in and create the unique "tones" or "resonances" that mathematically represent the different types of particles and their associated properties. More specifically, the number of dimensions required by superstring and M-theory arises from negative probabilities that arise in lower dimensional formulations when trying to incorporate quantum effects. With 9 and 10 spatial dimensions, the negative probabilities cancel out completely. The vibration resonance patterns of these strings give rise to the mass and charge of our elementary particles. The higher the frequencies in a string, the greater the mass of the elementary particle it will produce, and hence a lower vibration in a string would represent a lower mass particle. In addition to the unification powers inherent in the hyperspace constructs of string theories and M-Theory, the concept of a string in itself houses a powerful unification tool.

Prior to strings, theories conceived of particles mathematically as points of energy that had zero size. Mathematically this can cause problems under certain calculations as it can lead to potentially dividing by zero; which as we all remember from math class, is undefined as infinity. With strings, we never have to worry about dividing by zero, since the smallest constituent parts of the theory, the strings, have a definite—albeit small—size. This eliminates equations from ever dividing by zero, and thereby smoothes over the anomalies inherent in combining relativity with quantum physics. A potential problem for string theories is that the incredibly small size of strings means that they will likely never be detected directly experimentally. This leaves physicists searching for indirect evidence that strings and higher dimension spaces may leave behind, that is detectable. Clearly, regardless of how much elegance and power a theory is in unifying the forces of nature, if there is no way to validate it experimentally, then that theory is not science—it is philosophy.

 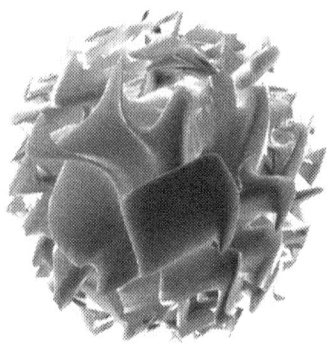

The degrees of freedom greatly increase with dimensionality. If we imagine these two objects as 2 dimensional and three dimensional strings, we can clearly see that by simply adding one more dimension, there are many more degrees of freedom for the string to move and vibrate. This concept is the part of the key to the success of string theories. Despite the rapid increase in the degrees of freedom with increased dimensionality, sting theory still requires ten spatial dimensions to account for all the observed particles and their behaviors.

Multiple Dimensional Spaces

The basic reason why so many dimensions are required by string theories is that when relativity and quantum physics were united via the concept of tiny vibrating strings of energy, the probabilities associated with quantum physics kept spitting out *negative* probabilities for events for lower dimensional spaces. It was not until calculations were tried in nine dimensional space (with one dimension for time, for a total of ten dimensions) that positive probabilities were achieved.

Ten or eleven dimensional space is not something that the human mind can fully comprehend. Our minds were born, raised, and trained to think and experience nature in terms of our three spatial dimensions and our one time dimension. As we explored previously in this book, it is important once again to emphasize what a dimension really is. Dimensionality is the unique degrees of freedom: *unique ways a system can move.* So how do we start to understand and visualize the concepts of higher dimensional space, like 5 dimensions? Let us once again use the power of analogy to help with this.

Imagine an immense canyon that has a bridge running across it. Our invented bridge is unique in that it is made entirely of a single huge piece of glass akin to taking a giant window pane and placing it over our canyon. However, this is no ordinary glass, as it only needs to be a millimeter thick. Let us now apply five dimensional space to our bridge by volunteering a person named Sam to walk explore the bridge and see how many unique ways she can move about on it. Before sending Sam out onto our glass pane bridge, we have her wear special shoes that have powerful suction cups on the soles that she can turn on and off whenever she wants. After much coaxing (after all, it is a glass bridge one millimeter thick), Sam finally starts exploring the bridge. From our vantage point on the side of the canyon, we can track all her movements. Since the bridge is at a fixed height, we can say that one of the dimensions is fixed (the z-value, altitude—dimension *one*). As Sam walks around, she first sees that she can walk all over the surface of the bridge without restrictions (dimension *two*). What else will Sam be able to do on our bridge? Over time, and after getting tired of walking around the surface of the bridge, Sam realizes that she can jump up off the surface of the bridge as well (dimension *three*). Without knowing, Sam has shown us another dimension of our bridge, that of *time*. It has been taking her time to walk around the bridge and jump about (dimension *four*). What about our elusive dimension five? For this dimension, we need Sam to go to edge of one of the bridges sides, and turn on her suction cup shoes. Again, after much contemplating, Sam finally works up the nerve to use her suction shoes (and a bit of gymnastics) to swing over the side of the bridge and let her shoes stick to the *bottom* of the bridge. Sam now has a whole new degree of freedom on this bottom surface with which she can walk around (dimension *five*). Since our bridge is glass, we can see Sam walking about on the bottom surface of the bridge. Let us suppose that we can change the opacity of our glass at a whim. At zero percent opacity we can see everything and the glass is perfectly clear. At 100% opacity, we are unable to see through the bridge at all. If we imagined another test subject, William, as walking around on the top surface of our bridge, while we had the opacity at 100%, he would not be able to see that Sam was walking around underneath him on her newly found dimension. To William on the surface of the bridge, Sam's dimension is *hidden*, despite her being right underneath him, one millimeter away (the glass's thickness). Unless Sam stomps her feet to send vibrations up to William, or yells out and William hears the echoes from the walls of the canyon, he cannot otherwise detect Sam's presence. As you can see with our analogy, with a little imagination, extra dimensions can be imagined, and the type of thinking in this analogy will go a long ways to helping your mind cope with extra dimensions. This is fine for five dimensions, however what about *eleven dimensional* space-time? Is there any way of imagining that?

Again, by the power of analogy, we can visualize even this complicated scenario using nothing but a hardware store nut, and a bit more imagination. While it may not seem like much of a constellation, we can eliminate the need to visualize one of these eleven dimensions as we only need to be concerned with the *spatial* dimensions, and not the dimension of time (a time dimension is pretty self-explanatory). Let us consider a regular octagonal hardware store nut. This may seem like a trivial type of object to use to visualize eleven dimensional spaces; however it turns out to work rather nicely. If we imagine scaling up our nut so that it is approximately the same size as our glass bridge from before, we can once again have our volunteer Sam, with her suction cup shoes, walk all over the surface and see how many unique ways she can move. Lets start…Sam first finds she can walk all over each of the eight sides of the nut, and that each of these sides gives her a certain unique range of motion as compared to the whole nut (dimensions 1 through 8). Next, Sam steps off the sides of the nut to the top of the nut (dimension 9). While walking on the top of the nut, she sees the circular hole bored through the nut, and steps down into it. Despite all of the threads getting in the way, she finds that if she stays in the "valleys" of the thread she circles around and eventually ends up at the other side (our bottom) of the nut (so the inside of the nut is dimension 10), where she hops out of the thread, and once again is free to walk around on (the back of the nut is dimension 11). While this example makes eleven dimensions seem trivial, there is a catch. In eleven dimensional space-time, *each and every possible point* in our three dimensional space would need to be represented as having such a "nut" (or comparable object—such as a Calabi-Yau shape—which will be explored later) tacked on to every point. If you take measurements of an event and graph it out, each point of data that you use to make the curve of your graph, when zoomed in, would be a separate "nut" in eleven dimensional space-time using our analogy. All events, from the paths traces out by atoms colliding in accelerators, to the motions of stars and planets, would be tracing out their events as strings of "nuts" in eleven dimensional spaces. Since the human mind cannot truly visualize so many dimensions as they would appear in reality, the object used, such as our nut, essentially becomes arbitrary, as there are many objects that could be viewed as eleven dimensional (or any number of dimensions) and used as equally valid examples. Technically, any multidimensional space that has more than three spatial dimensions is referred to as a *hyperspace* (the first person to explore higher dimensions was the mathematician Gunnar Nordström). Another popular way to imagine a tiny curled up dimension that is not visible is to think of a section of garden hose, where from a long distance away it appears to be a one dimensional line of no thickness, however since we know it is a garden hose, we know that at every point on the hose, there is a circular dimension that exists, we merely cannot see it from our distant vantage point.

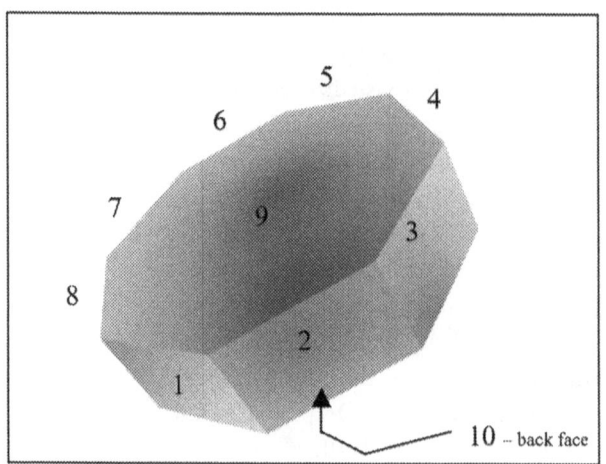

A simple representation of ten dimensional space. Each of the eight sides, and the front and back of the extruded octagon have unique degrees of freedom associated with them if you were a person walking on the surface of this object.

Calabi-Yau Shapes of Hidden Dimensions

It turns out that in the mathematics describing higher dimensions there are only certain types of shapes that will curl up in string theory to make hidden dimensions when the number of dimensions necessitated by modern string theories are taken into account. Just as any musical instrument is carefully designed to produce certain ranges and types of tones, so to was it found that the hidden spatial dimensions of string theories needed to be of particular types of shapes in order to provide strings with the right types of vibrations to account for the variety and properties of the observed subatomic particles. The mathematically viable shapes for these curled up spatial dimensions are called *Calabi-Yau shapes*. Geometrically, these shapes are akin to higher dimensional donuts or pretzels. If the higher dimensional spaces of string theory are proven experimentally, this means that these Calabi-Yau shapes are tacked on to every point in our three dimensional space. That would mean that every time we move, on the smallest of conceivable scales, our bodies are passing through these higher-dimensional shapes. While not an exact example, the figure on the following page approximates what a Calabi-Yau shape can look like:

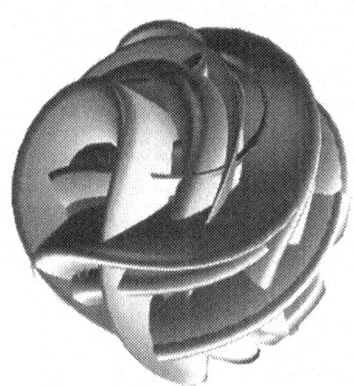

In the higher dimensional spaces required by string theories, such a shape would be tacked on at every point in our three dimensional space to house the hidden spatial dimensions that are curled up.

On the Cusp of Science and Philosophy

Despite superstring and M-theories tremendous unification powers, there are people who are critical of it. After all, necessitating the existence of six and seven extra spatial dimensions is a tremendous leap for any theory to make. Beyond this, there is the fact that physics is based on being able to compare theory against *experimental results*. The problem with strings is that they are so unimaginably tiny, that it is conceivable, and perhaps probable, that we will *never* be able to detect them directly. Strings are predicted to be on the order of a Planck length, or 1.6×10^{-35} meters, or 10^{-20} times the size of a proton. Obviously, at such tiny lengths our current realm of experiments cannot even come close to such scales. This leaves a puzzling remnant: If we are unable to experimentally test superstring and M-theory, then they cannot be considered a scientific theory and they will fall into the realm of philosophy. Why should there by any reason for anyone to believe a theory that is unable to be validated on any level by experiment? The bottom line is that you should not believe anything that cannot be experimentally proven. However here we are, talking about tiny vibrating strings of energy that may very well hold the key to the unification of all the forces of nature. That is because there are methods of *surmising or inferring* the properties of strings, and hidden or curled up dimensions (as in our second bridge example). A good example of inferring evidence is black holes. At this point in our understanding of physics, black holes are practically considered to be a proven fact. Despite this popular viewpoint, black holes have never been *directly* observed, and in all probability *never* will be observed directly. However, over the years, with evidence *inferred* from gravitational lensing and their associated Einstein rings and mirror images, and information from superheated, rapidly rotating accretion disks, we have surmised enough evidence about them that fits theory prediction from relativity to accept their existence in the scientific community.

Wormholes

By this point in this book, we have covered many profound and counter-intuitive concepts. Perhaps, at this point the notion of wormholes is not much of a leap. Most people associate worm holes with science fictions movies or television shows. However, in the realm of string theories, there is an allowance for this type of phenomena that is mathematically viable. In a nutshell, a wormhole is a tear in space-time that allows for a possible short-cut between two points in space. Einstein's theories of relativity allow for space to *stretch* in all sort of bizarre ways, however the mathematics of relativity do not allow for the *tearing or ripping* of the fabric of space. A wormhole is a tear in space-time itself which tunnels through it, and possibly, ends at another space-time "rip". It is the ultimate short-cut through space-time. We can picture this short-cut by means of example. Let us imagine a piece of paper as a huge piece of space-time. Let us now draw two dots on opposite ends of this piece of paper "space-time". We can picture these two dots being two distant points, such as separate galaxies or stars, which we wish to travel between. If we are bound to the "surface" of the paper, then obviously the quickest way to get from one dot to the other is to go in a straight line between them. We can bend, twist, or curl our paper space-time sheet in any way, and that straight-line distance between the points is constant. The trick of a wormhole is to imagine folding the paper in such a way that the dots are right over each other, but not quite touching. If we were *not* bound to the surface of the paper space-time, we could tear a hole in the space-time paper at one of the points, and simply hop across to the other dot, through a worm-hole, tear another hole in the paper space-time, and pop out on the other side. Depending on how close our dots are oriented based on how we bend and curl our space-time paper; we quickly see that the distance between two points using this method can be greatly reduced. This is the concept of a worm-hole.

Why are we wasting our time on this when earlier we mentioned how relativity does not allow for the ripping and tearing of space-time? We can still find this exploration profitable because string theory *allows* for rips and tears in space-time. In string theories, a closed-loop string traveling through space carves out a three-dimension "tube" through time. These "tubes" can surround rips and tears in space-time and allow the mathematics of the rip to be represented by a tube, instead of a single line (of zero thickness), thereby avoiding the infinities of trying to accommodate the same event in relativity.

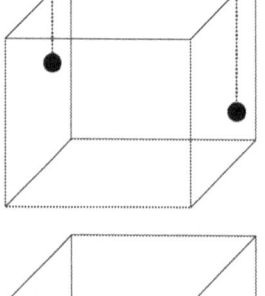

Without the presence of a wormhole, you are restricted to traveling on the surface of the space-time fabric, as the dotted path on this space-time cube portrays.

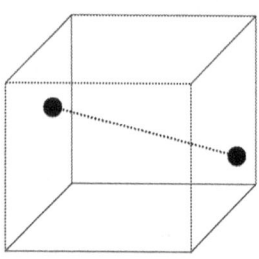

In the presence of a wormhole, there is a tear in the fabric of space-time itself, which enables you to tunnel through to a distant location much more easily, if they exist.

The Removal of Anomalies—The First Superstring Revolution

As with any new controversial theory, members of the physics and mathematics communities were reluctant to hop on board. As we explored earlier, in quantum theory there is the need to *renormalize* certain calculations to avoid infinities from plaguing the results. This renormalization was a mathematical trick of sorts that subtracted out the infinities from each side of equations to yield a finite solution that was in agreement with experimental results. While this technique is successful, it is sort of a mathematical band-aid. John Schwarz and Michael Green were two string theorists who were proponents of superstring theory—where the "super" in superstring stands for *super symmetric*. In a conference in Aspen Colorado, in 1984, after tedious calculations, Schwarz and Green announced that they had found superstring theory was totally *free of anomalies*. In other words, they proved mathematically that superstring theory that was free of infinities and violations of key conservations laws and hence did not require the trick of renormalization as other theories did. Ed Witten, a leading string theorist confirmed their results and gave them his seal of approval, which provided a much needed boost to the theory. This discovery and announcement by Schwarz and Green that superstring theory was anomaly-free is what is referred to as the *first superstring revolution*.

Strings Theories Galore

As with any revolution in theoretical physics, the first superstring revolution as unleashed by Schwarz and Green spawned a myriad of other viable theories to come forth. String theorists went from having no viable theories to having *five*, all within a year of Schwarz and Green's discovery of anomaly-free superstring theory. All five of these theories were variations on superstring theory, upon which vibrating strings of energy were the fundamental constituents. These theories ended up being named Type I, Type IIA, Type IIB, Heterotic-O and Heterotic-E (the four discoverer's of Heterotic string theories are all part of a Princeton team often referred to as the "Princeton string quartet"). Type I string theory allowed both open and closed string types in its framework, while Type IIA, IIB, and the Heterotic E and O theories only allow for closed strings.

The Introduction of the Graviton

Earlier, I mentioned the proposed carrier particle of the gravitational force—the graviton. In 1975, the Caltech Physicists John Schwarz, along with Joël Scherk discovered a connection between supersymmetry and gravity. More specifically, they discovered massless spin two bosons within the framework of string theory that were previously unknown. Other theorists had already known about these massless spin two bosons, however they did not apply any significance to them as they did not appear to match any of the observed known particles. Schwarz and Scherk did not take this path, instead they proposed that these massless spin two bosons were the force carrying particles of gravity—or *gravitons*. While a graviton has not ever been directly observed in a particle accelerator to date, there is much theoretical evidence to support its existence, and many researches consider it just a matter of time before a graviton is observed in the next generation of high energy accelerators being constructed.

The Second Superstring Revolution—The Birth of M-Theory

The initial surge in string theories brought about a unique kind of problem. Once all the dust settled, and there were five separate string theories that could house relativity and quantum physics under the same roof. There seemed to be "too many cooks in the kitchen". Physicists at first thought that after rigorous mathematical analysis, a fore-runner among these five string theories would victoriously emerge. However, as more time passed, and all five theories were still present, it became more of a concern. Clearly, some reducing was in order. We do not want to deal with five different "cooks" if they are all trying to bake the same thing. Dr. Ed Witten, a professor for the Princeton Institute for Advanced Study, noticed this theoretical surplus and decided to see if there was any way of unifying these separate string theories. What he came up with and announced at a 1995 String conference was the beginning of the M-theory, where the "M" stands for Mystery, Magic, Matrix or Membrane—based on your preference. M-theory can be imagined as a master theory that provides a way of connecting all of the viable string theories mentioned earlier. In essence, Witten made the realization that all five of the viable string theories were the looking at the same thing from different vantage points. In physicists' words, Witten had discovered the unifying *dualities* that existed among the separate string theories and provided a unified framework to incorporate all them within as *limits* of those dualities. Another way of thinking of M-theory is in terms of energy levels. It is thought that the five different string theories are merely the same concepts being studied at different energy levels, and M-Theory comes into play at energies that are high enough to include all of the lower energy levels of the other five string theories. Interestingly, the lowest energy level of M-theory reduces to eleven-dimensional supergravity theory. All the other superstring theories can be derived from supergravity theory via the rules set forth by M-Theory.

Part of the key to unify the string theories into M-theory was the addition of another dimension, making the grand total of eleven for space-time dimensions (three visible or "uncurled" dimensions, seven hidden or "curled" dimensions, and one time dimension). In traditional string theories the individual constituent strings were thought to be closed loops with no loose ends. With M-Theory, there is the possibility of open string with two ends, and these strings can be stretched out to form what are known as membranes, or *branes*. A brane is a two dimensional "sheet" of energy derived from the stretching out of a one dimensional open string. There is also the possibility of *p-branes*, which are essentially branes that have a higher dimension than two. Sometimes theorists will refer to branes by their dimensions as well—as a 2-brane, 3-brane, 4-brane…etcetera. The framework of M-theory allows branes of zero to nine dimensions. In the quest for the unification of the four forces of nature, M-theory is proving to be the leader of the pack. Ed Witten's discovery and announcement of M-theory in 1995 is referred to as the *second superstring revolution*.

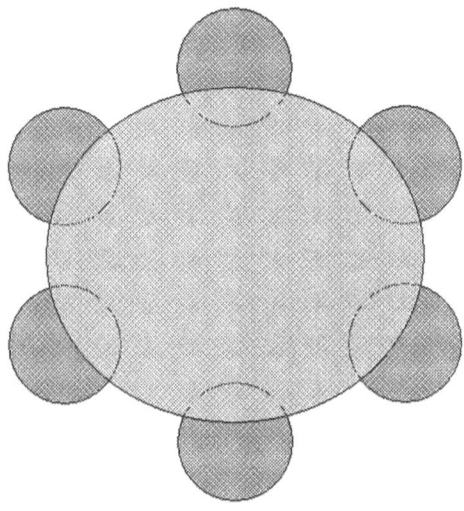

A visual representation of M-theory. The large center circle represents the framework for M-theory, where the smaller periphery circles represent the individual string theories and 11D supergravity. All of the different string theories and supergravity are merely different energy levels of the main M-theory.

Is Gravity Really Weak? A Possible Explanation

As we explored before, gravity is incredibly weak. In order to lift my fingers to type this section of the book for instance, I am easily overcoming the gravitational force of the *entire* planet earth. In light of developments in super-string, and M-theory, there is the possibility that gravity may in fact be just as strong as the other forces of nature. We may not be able to *feel* its full strength due to dissipation through hidden dimensions. In theories prior to superstring and M-theory, there were not any hidden dimensions where forces such as gravity's strength could "hide" or be diminished. In superstring and M-theory however, there are upwards of seven extra spatial dimensions that gravity could potentially influence. Specifically in M-theory, where open-ended strings can be stretched out into branes, the force carrying particle for gravity, the proposed graviton, could very well be a closed string with no ends. It could be that the strings that make up the other forces and our everyday matter particles are open-ended strings that have their ends "fastened" to the membrane that we call our four dimensional space-time. This means that open strings are limited in the sense that they can only influence other objects that are on the brane they are attached to. If the graviton were a closed string loop, then it would have no "ends" to anchor it and keep it from leaving the brane and allowing its presence (in the way of a gravitational force) to be felt in the other hidden dimensions, thereby reducing its apparent strength to us. If true, this would mean that when the force of gravity, as added up through all the extra dimensions and branes, could very well be every bit as powerful as the other forces of nature that are bound to our parent brane.

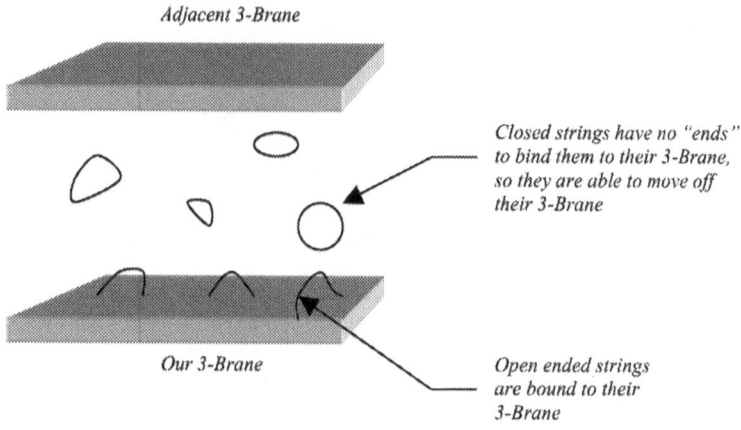

Adjacent 3-Brane

Closed strings have no "ends" to bind them to their 3-Brane, so they are able to move off their 3-Brane

Our 3-Brane

Open ended strings are bound to their 3-Brane

This concept of closed-loop gravity is of great interest to astronomers and cosmologists as it might be a possible explanation for the *dark matter* that is observed in galaxies and clusters of galaxies. This dark matter is thought to exist since the matter that we are able to detect in the electromagnetic spectrum is not enough to keep galaxies and clusters of galaxies together with their observed rotational velocities. Anyone who has played on a merry-go-round at a park knows that the faster you spin the merry-go-round, the more force it takes to keep you from flying off of it. This is the same for rotating galaxies and clusters of galaxies as well. The gravitational force that is present due to the mass of the detectable objects is not enough to keep such structures from flying apart. Therefore there must be some undetected *dark matter* present. The calculation of the quantity of this dark matter varies, however it is consistently over *90 percent*. This means that all the observable matter in our universe is, *at best,* 10 percent of the cosmos. In the most recent (and more accurate) calculations of dark matter percentages as calculated from WMAP data (Wilkinson Microwave Anisotropy Probe), the entire visible matter of the cosmos constitutes a mere 4% of the mass of the universe. If however, gravity is able to pass its influence between extra hidden dimensions, then it could be that this dark matter is nothing more than gravitons leaking off of other nearby dimensions of space.

Experimentally, such a theory of gravity passing from our dimension into others could be validated in high-energy colliders. In essence, the validity of the gravitons influence on other dimensions would be apparent in their absence (albeit this would be circumstantial evidence). If a graviton was detected and suddenly "vanished" soon after the collision, this could be an indirect way of determining that it left the "surface" of our brane into another dimension or brane. More specifically, we would not be able to detect the graviton itself; rather it is feasible to detect the predicted shower of hadrons (e.g. protons, neutrons, quarks) that would be emitted from particle accelerator collisions, thereby giving away the "presence" of the graviton before it vanished off to another dimension. While no such interaction has been seen as of this book's printing, the race is on to discover such an event. Higher energy colliders such as CERN's Large Hadron Collider (LHC) feasibly could be powerful enough to produce collision energies necessary to capture a graviton drifting off our brane.

Branes and the Big Bang

How does M-theory handle the big bang itself? Is there any way in which M-Theory can provide a plausible explanation for the big bang? It turns out that there are some possibilities. One that is intriguing is the possibility that the big bang was nothing more than the collision of two different branes; one which our known cosmos is confined to, and another brane (or branes) that are drifting around in higher dimensional spaces. Such a collision could feasibly produce the effects of a big bang on the surfaces of the branes. The problem with this approach is that theorists do not really know for sure mathematically what happens when such an event occurs. As such, it remains highly speculative, but a nonetheless engaging, possibility. What is even more interesting about this notion is that such a collision between branes could happen any number of times, even an infinite amount of times; making a big bang type of event a common event. This concept of colliding branes was originally introduced by Paul Steinhardt, Bert Ovrut, Neil Turok and Justin Khoury. In their original model, the branes were envisioned as being perfectly flat, thereby making a uniform exchange of energy across the branes when they collided. This would fit the observed uniformity that we know the known cosmos had in its youth based on the smoothness of the cosmic microwave background radiation that was emitted when the cosmos was only a few hundred-thousand year old. Prior to the advent of branes, this smoothness was explained by a brief, super-intensive *inflation* of space-time that occurred so rapidly as to smear out any irregularities in the first instants of the big bang across the whole universe. Even though this is a very intriguing possibility, most theorists are still on board with the inflationary variation to the big bang.

At the Cusp of Proof

With the new particle accelerators that are scheduled to be coming on line in the near future, there may not be as long a wait as one might think in mining out the validity of some of the key concepts of M-theory. The magnitude of energies needed to probe some of M-theory's predictions should be achieved in this next round of high energy colliders. Many people firmly believe that M-theory, or some variation of it, holds the key to the unification of all the forces in nature.

Concluding Remarks

While I have tried to touch base on as many ideas as possible in writing this book, it is truly impossible to keep up with all of the research and data that continually streams in from scientific fields. In our modern era, with computers and the internet, there is a staggering amount of data that can be produced from scientific studies in a short amount of time. Despite this, I hope that I have opened up a door in your mind that will allow the light of curiosity to flood in, and challenge you to learn more about the amazing directions that astronomy, physics, and cosmology are headed for in the upcoming decades. It has become common practice in modern science to unveil truths that were once considered science fiction.

Even with the vast technological and mathematical arsenal that modern science brings to bear, it is important not to forget that regardless how far science advances, it is always limited on a fundamental level to provide merely an approximation of true reality. To some, this may be a jagged pill of sorts; however to me personally, I find that maintaining a fundamental level of mystery in nature is the key to nurturing our thirst for knowledge. If we did ever in fact become all knowing of our environment, then there would cease to be a need for curiosity, as there would be no unknowns to be curious about.

While we may never be able to unravel the deepest secrets of nature, it is important to realize that we ourselves are the embodiment of the nature that we seek to understand through science. When we observe a distant galaxy, or a person crossing the street, or a gust of wind rustling some nearby leaves, we are in fact looking at ourselves in a very literal sense. All of the atoms that make up those galaxies are the same atoms that we are made of, and that the leaves are made of, and that the person crossing the street is made of. Every heavy atom that makes what we see around us possible was forged in the raging furnaces of the stars and supernova explosions. We are indeed made of "star stuff". We are all tied into the cosmos on an intrinsic level—we are the children of the cosmos. Everything is on its way to becoming part of something else. We are all a part of something *much* larger. Once again, I find myself thinking of Newton:

> "I have been like a boy, playing on the seashore, diverting myself and now and then finding a smoother pebble or prettier shell than usual, while the great ocean of truth lay before me, all undiscovered."

Bibliography

Books:

Richard P. Feynman, *QED* (Princeton University Press, 1985)

John Gribbin, *Schrödinger's Kittens and the Search for Reality* (Back Bay Books/ Little, Brown and Company, 1995)

James Gleick, *Chaos* (Penguin Books, 1987)

Leon M. Lederman, Christopher T. Hill, *Symmetry and the Beautiful Universe* (Prometheus Books, 2004)

Michio Kaku, *Parallel Worlds* (DoubleDay, 2005)

Tony Hey, Patrick Walters, *The New Quantum Universe* (Cambridge Press, 2003)

Paul Halpern, *The Great Beyond* (John Wiley & Sons, Inc., 2004)

Roger Penrose, *The Road to Reality* (Alfred A. Knopf/Random House, 2004)

Benoit B. Mandelbrot, *The Fractal Geometry of Nature* (1977)

Carl Sagan, *Cosmos* (Random House, 1980)

Eric Chaisson, Steve McMillan, *Astronomy Today* (Prentice Hall, Inc., 1996)

Michio Kaku, *Hyperspace* (Oxford University Press, 1994)

Jim Al-Khalili, *Quantum* (Weidenfeld & Nicholson, 2003)

Kenneth W. Ford, *The Quantum World* (Harvard University Press, 2004)

Bernard Schutz, *Gravity* (Cambridge University Press, 2003)

Albert Einstein, *Relativity* (Crown Publishers, 1961)

Richard P. Feynman, *Six Not-So-Easy Pieces* (Perseus Books, 1963)

L. E. Lewis, Jr., *Our Superstring Universe* (Universe, Inc, 2003)

Stephen W. Hawking, *A Brief History of Time* (Bantam Books, 1988)

Brian Greene, *The Elegant Universe* (Vintage Books, 1999)

John Gribbin, *The Search for Superstrings, Symmetry, and the Theory of Everything* (Back Bay Books/Little, Brown and Company, 1998)

Stephen W. Hawking, *The Theory of Everything* (New Millennium Press, 2002)

John D. Barrow, *The Book of Nothing* (Vintage Books, 2000)

Richard P. Feynman, *Six Easy Pieces* (Perseus Books, 1963)

Brian Greene, *The Fabric of the Cosmos* (Vintage Books, 2005)

Magazines:

Scientific American

Sky & Telescope

Web Sites:

http://archive.ncsa.uiuc.edu/Cyberia/NumRel/GenRelativity.html

http://particleadventure.org/particleadventure

http://particleadventure.org/particleadventure/frameless/chart.html

978-0-595-38706-9
0-595-38706-3

www.ingramcontent.com/pod-product-compliance
Lightning Source LLC
Chambersburg PA
CBHW030941180526
45163CB00002B/660